Advances in Experimental Medicine and Biology

More information about this series at http://www.springer.com/series/5584

Benjamin M.J Owens • Matthew A. Lakins

Editors

Stromal Immunology

 Springer

Editors
Benjamin M.J Owens
Somerville College
University of Oxford
Oxford, UK

EUSA Pharma
Hemel Hempstead
Hertfordshire, UK

Matthew A. Lakins
F-star Biotechnology Ltd.,
Babraham Research Campus
Cambridge, UK

ISSN 0065-2598 ISSN 2214-8019 (electronic)
Advances in Experimental Medicine and Biology
ISBN 978-3-030-08621-3 ISBN 978-3-319-78127-3 (eBook)
https://doi.org/10.1007/978-3-319-78127-3

Printed on acid-free paper

This Springer imprint is published by the registered company Springer Nature Switzerland AG.
The registered company address is: Gewerbestrasse 11, 6330 Cham, Switzerland

Preface

The concept for this book arose as a result of growing interest in the investigation of non-hematopoietic stromal cells and their impact on immune responses. Through interactions during diverse doctoral and postdoctoral research programmes spanning several years at the University of York, the University of Cambridge and the University of Oxford, we recognised a need for a cohesive group to bring together scientists interested in the concepts underlying stromal immunology, and the Stromal Immunology Group (StIG) was born.

Having organised several successful international StIG conferences, we felt that a missing part of the picture was an advanced book comprising a collection of writings from leaders in stromal immunology that could act as a primer for professional researchers new to this specialist field. This book would also provide support for the teaching of graduate and undergraduate students in science and medicine.

What follows is the collected work of scientists and physicians from across the world, all of whom share a belief in the huge potential for research into stromal immunology to contribute to medical research. Topics covered range from the interaction between leukocytes and lymph node stromal cells, inflammatory responses of mesenchymal stem cells and fibroblasts to the key roles of stromal cells in response to infection, the tumour microenvironment and the healthy and inflamed intestine. Important avenues for future research are addressed, as are the uses of advanced cell culture systems for the investigation of human tissue stromal cell function and stromal cell targeting for therapeutic benefit.

Numerous studies have addressed the significant therapeutic potential of exploiting stromal cells in combating disease. Pancreatic cancer, for example, and specifically pancreatic ductal adenocarcinoma (PDAC), is a stromal-rich, lethal malignancy fundamentally resistant to standard of care therapies. Much work has been carried out targeting the desmoplasmic nature of PDAC, particularly the cancer-associated fibroblast and endothelial cell containing component. Whilst strategies aimed at depleting these cell types to aid drug perfusion and immune cell infiltration work well in murine models, the translatability of such approaches remains in question.

This approach is not limited to pancreatic cancer. Many other stromal-rich tumours which employ a highly desmoplastic stroma as a physical barrier to immune cell infiltration could be treated in such a way. Breast, prostate, and colon cancer all recruit and influence their tumour microenvironment in order to regulate immune escape, promote metastasis and aid progression. These cancers, and more recently others such as non-small cell lung carcinoma, are put through a prognostic test evaluating their tumour:stroma ratio and the outcome is used to successfully predict prognosis and the chances of relapse. Soon, tools such as the tumour:stroma ratio measurement could serve as an influencing factor on suggested treatment and whether targeting the stroma is a valid approach for those specific diseases.

Similarly, gaining a deeper understanding of specific mediators of stromal cell activation in chronically inflamed tissue – such as the recent discovery of Oncostatin M as a driver of intestinal stromal cell activation during inflammatory bowel disease – may lead to the identification of novel therapeutic axes that can be targeted to revolutionise therapy for patients with these debilitating inflammatory conditions.

We hope that this introduction to advanced concepts in stromal immunology serves as a useful, stimulating and enjoyable tool for those with an interest in learning more about this exciting area of immunology, and we look forward to seeing the field expand and grow over the coming years.

And remember, *'It's all about the Stroma'*

Oxford, UK   Benjamin M.J Owens
Hertfordshire, UK
Cambridge, UK   Matthew A. Lakins

Acknowledgements

This textbook was produced as a collaboration between the British Society for Immunology, the BSI Stromal Immunology Group and Springer Publishing, USA. We are indebted to all the authors for their valuable contributions.

Contents

1 Leukocyte-Stromal Interactions Within Lymph Nodes............ 1
Joshua D'Rozario, David Roberts, Muath Suliman,
Konstantin Knoblich, and Anne Fletcher

2 Stromal Cell Responses in Infection 23
Paul M. Kaye

3 Fibroblasts and Osteoblasts in Inflammation and Bone Damage ... 37
Jason D. Turner, Amy J. Naylor, Christopher Buckley,
Andrew Filer, and Paul-Peter Tak

**4 Molecular and Cellular Requirements for the Assembly
of Tertiary Lymphoid Structures**............................ 55
C. G. Mueller, S. Nayar, J. Campos, and F. Barone

**5 Mesenchymal Stem Cells as Endogenous Regulators
of Inflammation**.. 73
Hafsa Munir, Lewis S. C. Ward, and Helen M. McGettrick

6 Stromal Cells in the Tumor Microenvironment.................. 99
Alice E. Denton, Edward W. Roberts, and Douglas T. Fearon

7 Immunosuppression by Intestinal Stromal Cells................. 115
Iryna V. Pinchuk and Don W. Powell

**8 Novel Models to Study Stromal Cell-Leukocyte
Interactions in Health and Disease** 131
Mattias Svensson and Puran Chen

Index.. 147

Contributors

F. Barone Rheumatology Research Group, Institute of Inflammation and Ageing (IIA), University of Birmingham, Birmingham, UK

Christopher Buckley Rheumatology Research Group, Institute for Inflammation and Ageing, College of Medical and Dental Sciences, University of Birmingham, Queen Elizabeth Hospital, Birmingham, UK

The Kennedy Institute of Rheumatology, University of Oxford, Oxford, UK

J. Campos Rheumatology Research Group, Institute of Inflammation and Ageing (IIA), University of Birmingham, Birmingham, UK

Puran Chen Center for Infectious Medicine, F59, Department of Medicine, Karolinska Institutet, Karolinska University Hospital, Huddinge, Stockholm, Sweden

Joshua D'Rozario Department of Biochemistry and Molecular Biology, Monash University, Clayton, VIC, Australia

Institute of Immunology and Immunotherapy, University of Birmingham, Birmingham, UK

Alice E. Denton Lymphocyte Signalling and Development, Babraham Institute, Cambridge, UK

Douglas T. Fearon Cold Spring Harbor Laboratory, Weill Cornell Medical College, New York, NY, USA

Andrew Filer Rheumatology Research Group, Institute for Inflammation and Ageing, College of Medical and Dental Sciences, University of Birmingham, Queen Elizabeth Hospital, Birmingham, UK

Anne Fletcher Department of Biochemistry and Molecular Biology, Monash University, Clayton, VIC, Australia

Institute of Immunology and Immunotherapy, University of Birmingham, Birmingham, UK

Paul M. Kaye Centre for Immunology and Infection, Department of Biology and Hull York Medical School, University of York, York, UK

Konstantin Knoblich Department of Biochemistry and Molecular Biology, Monash University, Clayton, VIC, Australia

Institute of Immunology and Immunotherapy, University of Birmingham, Birmingham, UK

Helen M. McGettrick Rheumatology Research Group, Institute of Inflammation and Ageing, University of Birmingham, Birmingham, UK

C. G. Mueller CNRS UPR 3572, Laboratory of Immunopathology and Therapeutic Chemistry/Laboratory of Excellence MEDALIS, Institut de Biologie Moléculaire et Cellulaire, Université de Strasbourg, Strasbourg, France

Hafsa Munir MRC Cancer Unit/Hutchison, University of Cambridge, Cambridge, UK

S. Nayar Rheumatology Research Group, Institute of Inflammation and Ageing (IIA), University of Birmingham, Birmingham, UK

Amy J. Naylor Rheumatology Research Group, Institute for Inflammation and Ageing, College of Medical and Dental Sciences, University of Birmingham, Queen Elizabeth Hospital, Birmingham, UK

Iryna V. Pinchuk Departments of Internal Medicine, University of Texas Medical Branch, Galveston, TX, USA

Microbiology and Immunology, University of Texas Medical Branch, Galveston, TX, USA

Don W. Powell Departments of Internal Medicine, University of Texas Medical Branch, Galveston, TX, USA

Neuroscience, Cell Biology and Anatomy, University of Texas Medical Branch, Galveston, TX, USA

David Roberts Institute of Immunology and Immunotherapy, University of Birmingham, Birmingham, UK

Edward W. Roberts Department of Pathology, University of California San Francisco, San Francisco, CA, USA

Muath Suliman Institute of Immunology and Immunotherapy, University of Birmingham, Birmingham, UK

Mattias Svensson Center for Infectious Medicine, F59, Department of Medicine, Karolinska Institutet, Karolinska University Hospital, Huddinge, Stockholm, Sweden

Paul-Peter Tak Division of Clinical Immunology & Rheumatology, Academic Medical Center/University of Amsterdam, Amsterdam, The Netherlands

Jason D. Turner Rheumatology Research Group, Institute for Inflammation and Ageing, College of Medical and Dental Sciences, University of Birmingham, Queen Elizabeth Hospital, Birmingham, UK

Lewis S. C. Ward Discovery Sciences, AstraZeneca, Cambridge, UK

Chapter 1
Leukocyte-Stromal Interactions Within Lymph Nodes

Joshua D'Rozario, David Roberts, Muath Suliman, Konstantin Knoblich, and Anne Fletcher

Abstract Lymph nodes play a crucial role in the formation and initiation of immune responses, allowing lymphocytes to efficiently scan for foreign antigens and serving as rendezvous points for leukocyte-antigen interactions. Here we describe the major stromal subsets found in lymph nodes, including fibroblastic reticular cells, lymphatic endothelial cells, blood endothelial cells, marginal reticular cells, follicular dendritic cells and other poorly defined subsets such as integrin alpha-7+ pericytes. We focus on biomedically relevant interactions with T cells, B cells and dendritic cells, describing pro-survival mechanisms of support for these cells, promotion of their migration and tolerance-inducing mechanisms that help keep the body free of autoimmune-mediated damage.

Keywords Stromal cells · Lymph nodes · Fibroblasts · FRCs · Lymphoid fibroblasts · Lymphatic endothelium · Endothelial cells · LECs · Stromal Immunology · Podoplanin · Non-haematopoietic

1.1 Introduction

Lymph nodes are the most prevalent secondary lymphoid organ (SLO), contained in the neck, armpits, lungs, abdomen, collarbone, knee and groin regions [1]. They range in size from a few millimetres to over 2 cm and enlarge significantly under certain conditions involving immune activation, such as infection or cancer [1, 2].

J. D'Rozario · K. Knoblich (✉) · A. Fletcher (✉)
Department of Biochemistry and Molecular Biology, Monash University, Clayton, VIC, Australia

Institute of Immunology and Immunotherapy, University of Birmingham, Birmingham B15 2TT, United Kingdom
e-mail: konstantin.knoblich@monash.edu; anne.l.fletcher@monash.edu

D. Roberts · M. Suliman
Institute of Immunology and Immunotherapy, University of Birmingham, Birmingham B15 2TT, United Kingdom

Lymph nodes are structurally organised and contain a cortex, paracortex and medulla, which are separated into different regions to allow the movement of lymph through the organ [3]. The cortex is situated beneath the capsule and subcapsular sinus with B lymphocytes and follicular dendritic cells contained within follicles present in the cortical region [4]. The paracortex lies deeper within the lymph node structure with T lymphocytes homing to these regions to interact with antigen-presenting cells [4]. The medulla consists of B lymphocytes and macrophages dispersed within medullary cords which allow for the movement of lymph from the cortex into efferent lymphatic vessels [4]. This structure allows for antigen-bearing antigen-presenting cells (APCs) and lymphocytes to efficiently interact within the lymph node, enabling an appropriate immune response against an invading pathogen [5]. The microenvironment of the lymph node is crucial for immune function and consists of endothelial cells lining lymphatics and blood vessels and fibroblastic reticular cells which create the internal reticular structure of lymphoid organs [6].

Lymph, which may bear soluble antigen, enters the lymph node through afferent lymphatic vessels, where it empties into the subcapsular sinus and then traverses through the medullary sinuses surrounding the medullary cords to interact with B cells [4]. The lymph filters though the cortex where it exits via the efferent lymphatic vessels contained in the hilus [4]. Dendritic cells (DCs) actively migrate into the lymph node via afferent lymphatics [7]. Dendritic cells then migrate to paracortical T cell zone using stromal cells as a scaffold for migration [8]. B and T cells also use the stroma to migrate, entering lymph nodes from the bloodstream through specialised high endothelial venules [9]. Following entry, B cells move to the B cell follicles in the cortex, while T cells move to the paracortical T cell zone where they can begin scanning arriving APCs for their cognate antigen [8, 10, 11].

Structural components of the lymph node are now broadly appreciated as primary regulators of the adaptive immune response [10, 12–26]. These lymphatic structural components, termed lymph node stromal cells (LNSCs), comprise of non-haematopoietic cells that can be divided into functionally and phenotypically distinct subsets based on surface expression of glycoproteins CD31 and podoplanin (gp38) with an absence of haematopoietic marker CD45 [14]. These include blood endothelial cells (BECs), lymphatic endothelial cells (LECs), integrin α7+ pericytes (IAPs), follicular dendritic cells (FDCs) and fibroblastic reticular cells (FRCs) [14, 22, 27]. These stromal cells play a variety of roles in lymph node homeostasis and function, as they interact with lymphocytes to create an optimal microenvironment for cell activation and migration.

1.2   Fibroblastic Reticular Cells (FRCs)

Selectable markers: Gp38+, CD31-, ER-TR7+, LTβR+, desmin+, aSMA+

FRCs are myofibroblasts that have evolved to create a specialised microenvironment within lymph nodes. FRCs are heterogeneous and exist in different niches within the lymph node, fulfilling unique immunoregulatory roles [27] (Table 1.1, Fig. 1.1a–f).

Table 1.1 Lymph node FRC subsets

FRC subtype	Characteristics	Phenotype	Function	References
T cell zone reticular cells (TRCs)	Secretion of CCL19,CCL21 and IL-7 within paracortex	PDPN+, desmin+, MAdCAM-, CCL19+, CCL21+	Maintaining T cell homeostasis Forming conduit network Allowing lymphocytes to migrate and interact efficiently on the 3D meshwork	[8, 11, 14, 22, 30, 33, 36]
B cell zone reticular cells (BRCs)	Located in (resident) or near (inducible) B cell follicles. They secrete BAFF and are induced during inflammation to produce CXCL13	Resident cells: PDPN+, CCL19+, BAFF+ Inducible cells: PDPN+, CXCL13+	Supporting B cell survival and follicle boundary integrity	[24, 25, 37, 38]
Marginal reticular cells (MRCs)	Located in subcapsular region Not found in tertiary lymphoid organs	PDPN+, desmin, MADCAM1, IL-7^{Hi}, CXCL13+, RANKL^{Hi}	Produce very high levels of IL-7. Precursor cell type for FDCs within lymph nodes	[39–41]
Follicular dendritic cells (FDCs)	Within lymph nodes, FDCs develop from MRCs but are nonetheless highly distinct from other FRC types. Located within primary and secondary B cell follicles. Secretion of CXCL13	CD21+, CD35+, MFGE8+, CXCL13+, ICAM1+, VCAM1, BAFF+	Maintains germinal centre integrity. Facilitates the production of high-affinity antibodies	[41, 42]
Pericytic FRCs	Surrounds HEVs PDPN signals to CLEC-2 on platelets to maintain endothelial integrity	PDPN+	Prevents bleeding from HEVs into lymph nodes	[34]
Medullary FRCs	Associated with plasma cells and macrophages	PDPN+	Poorly studied	[4]

1.2.1 Structural Roles

FRCs play crucial roles in secreting extracellular matrix components and forming a cellular meshwork to give the lymph node strength, flexibility and structure [8, 14, 22, 28–30].

While not a focus of this review, as a general characteristic, FRCs secrete a broad array of extracellular matrix components, including collagens and laminins, decorin, biglycan, fibromodulin and vitronectin, to maintain the lymph node structure [8, 14, 22] (Fig. 1.1d).

T zone resident FRCs facilitate leukocyte migration and priming by supporting and secreting a 3D conduit network to maintain the lymph node microenvironment [8, 14, 30–32]. Conduits are microtubules created by FRCs, which secrete constituent

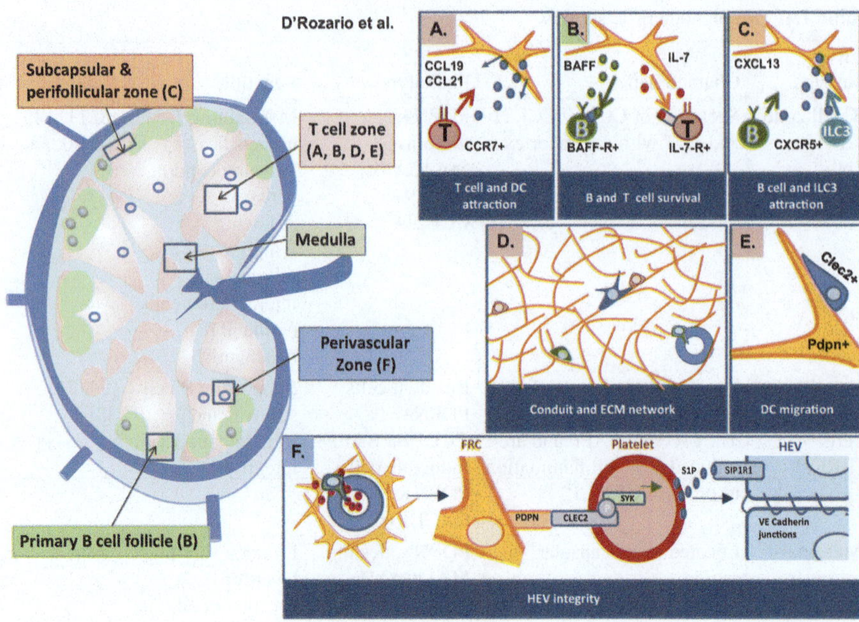

Fig. 1.1 FRC subsets reside in different lymph node niches and fulfil distinct functions. (**a**) T zone FRCs produce CCR7 ligands CCL19 and CCL21, which promote migration of naïve T cells and dendritic cells [8, 11]. (**b**) Within primary B cell follicles, B zone FRCs produce BAFF to promote the survival of naïve B cells [24], while within the paracortex, T zone FRCs produce IL-7 to promote the survival of naïve T cells [14]. (**c**) Within the subcapsular zone, marginal reticular cells produce CXCL13 to interact with innate lymphoid-like cells [39]. In the perifollicular zone close to primary B cell follicles, during inflammation some FRCs inducibly express CXCL13 to facilitate B cell follicle expansion [37]. (**d**) FRCs within the T zone create the conduit network through secretion of basement membrane and other extracellular matrix components [8, 14, 22, 33]. (**e**) T zone FRCs drive DC migration via signalling through DC-expressed CLEC-2, which binds podoplanin expressed by FRCs [56]. (**f**) In the perivascular zone, FRCs maintain the blood-lymph barrier by responding to infiltrating platelets. Platelets express CLEC-2, which binds podoplanin on FRCs, delivering a SYK-mediated signalling cascade that results in release of sphingosine-1-phosphate-1 from platelet surfaces, which binds S1P1-receptor on high endothelial venules, stimulating upregulation of VE-cadherin, which tightens endothelial cell junctions and prevents further nontargeted cell and liquid influx from the bloodstream [34]

basement membrane components and ensheath them. They permit low-molecular-weight molecules arriving via lymphatics to permeate quickly into the T cell zone to access resident dendritic cells (DCs) [30, 32, 33]. This allows DCs to rapidly process and present antigen to scanning T cells, permitting speedy initiation of an adaptive immune response. FRCs also surround high endothelial venules (HEVs) where they maintain the blood-lymph barrier by signalling to CLEC2 expressed by infiltrating blood-borne platelets, via the FRC-expressed glycoprotein ligand podoplanin [34]. CLEC-2 signalling induces the release of sphingosine-1-phosphate from the platelet surface, which regulates the binding strength of endothelial cells through VE-cadherin junctions [34].

In response to infection or inflammation, FRCs dynamically regulate lymph node expansion and contraction through expression of podoplanin, which maintains actomyosin contractility under homeostatic conditions and permits relaxation when it binds its ligand CLEC-2, expressed by DCs during inflammation [28, 29]. FRCs also proliferate during inflammation [28, 29, 35]. These dual mechanisms allow the lymph node to dynamically respond to and accommodate changing numbers of lymphocytes during activation and contraction phases of the immune response [28, 29, 35].

These important structural roles for FRCs are briefly discussed here but have been reviewed in detail elsewhere [22, 27].

1.2.2 Interactions with T Cells

1.2.2.1 Provision of Migration and Survival Factors

FRCs exist throughout the paracortical T cell zone; accordingly, interactions with T cells have been most closely studied. Chemotactic factors secreted from paracortical FRCs create the T cell zone by attracting naïve T cells and antigen-presenting cells (APCs) allowing them to initiate an immune response [10, 11, 14, 43]. This occurs through secretion of CCL19 and CCL21, which signal to CCR7 expressed by naïve T cells, leading to their migration through the lymph node [8, 10, 43, 44] (Fig. 1.1a).

T zone FRCs have also been shown to promote the survival and turnover of naïve lymphocytes via the secretion of T cell survival and growth factor IL-7 [14] (Fig. 1.1b). The secretion of this factor regulates and maintains the naïve CD4+ and CD8+ T lymphocyte pool within the lymph node ready for cell priming [14].

These effects of FRCs are particularly relevant to naïve T cells, since activated T lymphocytes within the lymph node are retained and can continue to function when FRCs are depleted [25].

1.2.2.2 Suppressive Tolerance

FRCs are capable of robustly suppressing CD8+ T cell proliferation early after their activation [19–21] (Fig. 1.2a).

Early after activation, T cells secrete inflammatory cytokines interferon gamma (IFN-γ) and tumour necrosis factor alpha (TNF-α), which stimulate FRCs to secrete nitric oxide (NO). NO acts in a paracrine manner on T lymphocytes curbing their proliferation [19–21]. NO is a highly pleiotropic molecule which facilitates many metabolic and immunologic pathways within the body [45]. Activated T cell-derived factors increase NO-producing enzyme nitric oxide synthase 2 (NOS2) mRNA and protein levels in FRCs leading to the release of NO [19–21]. Accordingly, NOS2−/− FRCs are unable to mediate T cell suppression [19–21]. Cyclooxygenases 1 and 2

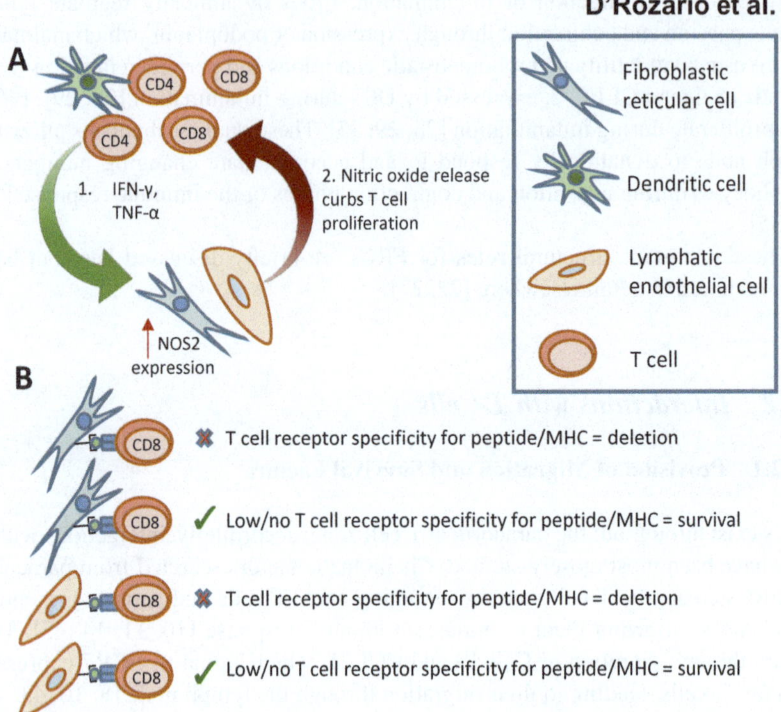

Fig. 1.2 Lymph node stromal cells impose suppressive and deletional tolerance. (**a**) Newly activated CD8+ T cells produce IFN-γ, TNF-α and an unidentified signal, which induce FRCs to increase expression of enzyme nitric oxide synthase 2 (NOS2) and produce nitric oxide (NO). FRC-derived NO acts on the T cell population to curb proliferation [19–21]. LECs are similarly capable of releasing NO to curb T cell proliferation [20]. (**b**) LECs and FRCs present endogenously expressed tissue-restricted antigens on MHC class I molecules to CD8+ T cells and delete T cells that respond with sufficient affinity [17, 18]. Similar mechanisms involving MHC class II-dependent antigen presentation are likely to operate for CD4 T cells [23, 26]

(COX1 and 2) in conjunction with NO2 expression have been hypothesised to play a potential role in T cell suppression, though further study is required [21].

This mechanism functions in vivo [20], though the immunological reach is still poorly understood.

Lymph node stromal cells may also induce tolerance in CD4+ T cells types through expression of MHC class II and associated antigen presentation pathway molecules under steady state and inflammatory conditions [22, 23, 26]. This theory has been reinforced by the ability of lymph node stromal cells to tolerise CD4+ T cells through the presentation of self-antigens via peptide-MHC class II expression [23, 26] and to induce homeostatic Treg proliferation [23]. In vitro data suggest that FRCs, LECs and BECs may acquire MHC II molecules from dendritic cells through cell-to-cell contact [26].

1.2.2.3 Deletional Tolerance

T cell tolerance induction by lymphoid stromal cells was first noted by Lee et al. [13], who showed that CD8+ T cells expressing a TCR reactive to ovalbumin (OVA) were specifically tolerised following interactions with lymph node stromal cells expressing OVA and that this prevented mice that expressed OVA in the gut (iFABP-tOVA) from developing autoimmunity. It was then shown that this response was due to FRCs and that FRCs could directly present self-antigen to T cells via MHC class I [18] (Fig. 1.2b), demonstrating that FRCs were capable of deleting autoreactive T cells and preventing autoimmunity. PD-1−/− T cells or PD-L1-blocking antibodies have been used to interfere with tolerance mechanisms in the iFABP-tOVA model, causing autoimmune enteritis [46].

FRCs were shown to express an array of organ-specific and tissue-restricted antigens [17, 18]. Tissue-restricted antigens (TRAs) are self-antigens native to peripheral tissues and organs and expressed at low levels within lymphoid organs for the purpose of educating the immune system for tolerance induction [47]. A major regulator of TRA expression in the thymus is the autoimmune regulator gene (Aire) [47]. However, in non-haematopoietic lymph node stromal subsets, Aire is not expressed [18]. It has been shown in human and murine tissues that increased expression of the Aire-like *protein DEAF-1* correlates with peripheral tissue antigen (PTA) expression [48]. While other factors may simultaneously exist, these results suggest a role of the *DEAF-1* gene in regulating lymphatic PTA expression, which requires further elucidation.

1.2.2.4 Systemic Effects of Interactions with T Cells

The depletion of FRCs has been shown to significantly attenuate cell-mediated immunity, as FRCs are required for the initiation of antiviral immune responses [25, 49]. In conditional FRC knockout models (DM2 BAC transgenic/FAP-DTR mice), naïve T and B lymphocytes were significantly depleted resulting in poor T and B cell-mediated immune responses during influenza A virus infection [25].

Similarly, the CCL19-Cre × Ltbr$^{-/-}$ mouse, which has an abnormal FRC network low in podoplanin, CCL19, CCL21 and IL-7, was unable to clear systemic lymphocytic choriomeningitis virus (LCMV) or mouse hepatitis virus showing a requirement for full FRC maturity [49]. These mice showed a 60–70% depletion of T cells and were unable to clear the viral infections by day 10 compared to control mice [49].

CCL19-Cre × iDTR mice, which are susceptible to inducible depletion of FRCs upon administration of diphtheria toxin, exhibited the loss of naïve CD4+ and CD8+ T cells within the lymph node during FRC ablation, as immunisation of mice with inactive influenza A virus led to an impairment of T cell priming and proliferation, with deterioration of antiviral T cell responses [24].

Furthermore, transplantation of IL-7Cre × R26-EYFP mice lymph nodes into C57BL/6 mice have shown that FRCs play a crucial role in providing IL-7 to initialise successful reformation of lymph node structure after avascular transplantation [50]. IL-7 derived from FRCs was also shown to promote T cell immunocompetence leading to structural adaptation of the lymph node microenvironment following systemic viral infection [50]. Damage to FRCs in clinical settings, in particular HIV infection, causes profound T cell immunodeficiency independent of viral load [51, 52].

1.2.3 Interactions with B Cells

FRCs in primary B cell follicles produce B cell-activating factor (BAFF) [24], a cytokine which drives the proliferation and maturation of B cells [53]. The production of this cytokine within primary follicles provides a favourable niche for B lymphocytes to develop [24] (Fig. 1.1b). Accordingly, FRC depletion has been shown to reduce the pool of naive B cells within lymph nodes [24, 25].

FRCs in the perifollicular zone have been shown to produce CXCL13 during infection, enabling the B cell follicle to expand and provide a favourable microenvironment for B cell activation and maturation [37, 54] (Fig. 1.1c). Inflammation was initiated by the injection of complete Freund's adjuvant into the ears of mice, and interactions with B cell zone reticular cells were analysed in ear-draining lymph nodes [37]. During systemic inflammation, B cells entered T zone areas of the lymph node in response to CXCL13 expressing B cell zone reticular cells, to expand the B follicle region [37].

1.2.3.1 Systemic Effects of Interactions with B Cells

Cremasco et al. [24] portray the loss of FRCs as detrimental to humoral immunity with immunisation with an inactivated influenza A virus leading to a reduction in influenza-specific immunoglobulin M in conditional FRC knockouts (CCL19-Cre × iDTR$^{fl/fl}$ mice) compared to control mice. In addition, these mice also exhibited impaired B cell viability and poor B cell follicle organisation, suggesting a systemic FRC importance in humoral immune responses.

In mouse graft-versus-host disease (GVHD) models, CD157+ FRC damage has been shown to impair IgG and IgA humoral immune responses to subcutaneous and oral antigens as B cell follicles are disrupted following FRC reduction [55].

1.2.4 Interactions with Dendritic Cells

Lymph node stroma has been shown in vivo to promote DC motility into and within the lymph node via the interactions between FRCs or LECs bearing podoplanin and activated DCs expressing Clec-2 [56]. Using Clec1b (CLEC-2) $^{-/-}$foetal liver chimeras compared to wild type, it was shown that CLEC-2+ DCs navigate from parenchymal tissues to lymphoid organs by migrating along stromal scaffolds that display the glycoprotein podoplanin [56]. Activation of CLEC-2 by podoplanin downregulates RhoA activity and phosphorylation of myosin light chains, causing cell spreading, and induces formation of protrusions through Vav signalling and activation of Rac1 [56]. Together these mechanisms promote DC motility across LEC and FRC stromal surfaces to allow antigen-bearing DCs to reach the lymph node and migrate within it in search of antigen-specific T cells.

Recent work highlights an important role for DCs in maintaining FRC survival and proliferation. DCs directly maintain FRC survival through provision of lymphotoxin ligands, which bind lymphotoxin beta receptor (LTbR) on FRCs, which upregulates podoplanin, in turn providing survival stimulus through maintenance of integrin-mediated adhesions [57]. Chyou et al. [58] showed that the initiation of FRC proliferation early in infection does not require DCs but that DCs induce FRCs to upregulate VEGF, which drives expansion of BECs and LECs. Yang et al. [35] revealed that DCs play an important indirect role in initiating FRC proliferation, by inducing naïve lymphocyte trapping within the lymph node early after infection is sensed. Moreover, various chains of MHC class II molecules were shown to become upregulated under inflammation on LECs, FRCs and BECs [22, 23, 26]. This suggests that subsets of LNSCs may be transcribing MHC class II molecules and/or receiving peptides from antigen-presenting cells, demonstrating a further encompassing role of LNSCs in innate immune responses.

1.2.5 Direct Detection of Inflammatory Stimuli and Interactions with Other Immune Cells

FRCs may be involved in the detection of lymph-borne infection or inflammatory signals via the expression of genes associated with pattern recognition toll-like receptors (TLRs) 3 and 4 [18, 22]. As TLRs respond to foreign pathogens by alerting the immune system, this data suggests that FRCs may directly detect viruses and bacteria. This idea has been reinforced by various studies which have documented the (direct or indirect) activation of LECs and FRCs via the usage of viral and bacterial immunostimulants or analogues which interact with TLR 3 and TLR 4 [18, 22, 28, 29, 35, 59].

The upregulation of chemoattractants and regulatory factors associated with the innate immune response has also been identified by transcriptional analysis of LNSCs [22]. FRCs express CXCL1, CXCL10, CCL2, CCL7, IL-33, IL-34, CSF1, CCL5 and CXCL9 and also express receptors for type I and II interferons [22].

1.3 Marginal Reticular Cells

Marginal reticular cells (MRCs) populate the outer regions of the cortex of lymph nodes [39]. They are located deep to the floor of the subcapsular sinus (SCS) and are phenotypically distinct from T zone fibroblastic reticular cells (FRCs) and follicular dendritic cells (FDCs). MRCs strongly express MAdCAM-1, CXCL13 and RANKL [39, 40]. The latter is an essential cytokine for lymph node development [60, 61]. However, the markers CCL21 expressed by T zone FRCs and CR1/CD35 expressed by FDCs are, respectively, absent or only trace expressed [39], indicating that MRCs are indeed a distinct stromal subset to these populations. Phenotypically similar groups of reticular cells have been found in other secondary lymphoid organs (SLOs), including the spleen and mucosa-associated lymphoid tissues [39, 62]. Contrastingly no similar groups of cells have been found in the tertiary lymphoid organs (TLOs) [5] associated with chronic inflammation.

During organogenesis, lymph nodes develop from accumulations of mesenchyme and haematopoietic cells associated with epithelium or vasculature, known as anlagen [63]. The haematopoietic cells are known as lymphoid tissue inducer (LTi) cells which bear the phenotype CD45+ CD4+ CD3−. LTi cells interact with the mesenchymal cells known as lymphoid tissue organiser (LTo) cells. LTo cells express adhesion molecule (ICAM-1, VCAM-1, MAdCAM-1) and chemokine (CXCL13, CCL19, CCL21) profiles upon stimulation by LTi through their secretion of lymphotoxin (LT)-$\alpha 1\beta 2$ [63, 64]. Subsequently CXCL13 attracts LTi through binding cells at its CXCR5 receptor, propagating a positive feedback loop of development [63, 65–67]. MRCs are thought to be a direct descendent of LTo cells [40]. While yet to be proven, supportive evidence includes their similar molecular phenotypes and the high concentration of LTo cells and RANKL expression in the outer areas of embryonic LNs, which in adult lymph nodes becomes a niche for MRCs [39, 68].

In addition to their embryonic developmental role, the ability of MRCs to give rise to FDCs in the adult lymph node has also been demonstrated. The MRCs exhibit maturation into a transitional phenotype before evolving into phenotypically mature FDCs through a two-step process [41]. FDCs in the spleen arise through other mechanisms, developing from perivascular precursors [69].

Immune-stromal interactions of MRCs are still poorly understood. MRCs are located in various SLOs adjacent to the primary route of antigenic entry, suggesting that they may play a role in the regulation of antigen transportation pathways [39, 68]. In adult mice, MRCs produce CXCL13 to attract CXCR5+ innate lymphoid-like cells type 3 (ILC3), which drive lymph node repair and regeneration after damage [50, 70]. CXCL13 is also a B cell chemoattractant, and it has been hypothesised that MRCs may be involved in the transport of antigens from the SCS into the B follicle or to facilitate the motility of B cells in the outer follicle through their expression of adhesion molecules [39].

The cytokines IL-7 and RANKL are both expressed to high levels by MRCs [39, 68] and are thought to be crucial for lymphoid homeostasis. IL-7 is a naïve T cell survival factor suggesting that it is also involved in regulation of T cell survival. In an in vitro murine model, the inoculation of mice with LTβR-Fc resulted in disorganisation of the follicular assembly of the white splenic pulp, and the disappearance of MRC layers, demonstrated by a loss of CXCL13 and RANKL staining [68]. In lymph nodes, LTβR-Fc downregulated CXCL13 but did not alter RANKL staining. This indicates that either RANKL expression in lymph nodes is independent of LTβR-NIK signalling or that another RANKL-expressing cell type is able to compensate for loss of LTβR ligands within the lymph node [68].

Their anatomical placement proximal to the inflow tract of antigens in SLOs, along with cytokine and chemokine expression, suggests that MRCs have an important role in regulating lymphoid function and SLO homeostasis.

1.4 Lymphatic Endothelial Cells (LECs)

Selectable markers: Gp38+, CD31+, Lyve-1

1.4.1 Structural and Chemoattractive Role

LECs create afferent and efferent lymphatic vessels, primarily to allow for the entry of antigen-presenting dendritic cells and soluble antigens into the paracortex of the lymph node [71, 72] and the egress of lymphocytes from the medulla [73]. LECs are also contained within medullary sinuses and line the ceiling (cLECs) and the floor (fLECs) of the subcapsular sinus [72]. It is thought that because of their prime position close to lymph, LECs might also be an early cell type to encounter and present antigens by environmental sampling [74]. LECs from different areas of the lymph node show differing expression of key surface markers: subcapsular LECs are PD-L1hi, ICAM-1hi, MAdCAM-1+ and LTbRlo; medullary LECs are PD-L1hi, ICAM-1hi, MAdCAM-1neg and LTbR+; and cortical LECs are PD-L1int, ICAM-1int, MAdCAM-1neg and LTbR+ [75].

Under inflammatory conditions, LECs direct macrophages and antigen-bearing dendritic cells along lymphatics, between LECs lining the subcapsular sinus, into the lymph node structure. LECs express podoplanin and similar to FRCs are capable of driving DC migration towards and within lymph nodes through signalling to CLEC-2 [56] (see Sect. 1.2.4). They produce CCL21 [22] and are thought to regulate availability of chemokines such as CCL21 and CCL19 in the subcapsular sinus and local parenchymal tissue through expression of scavenging receptors ACKR2 and ACKR4 [76] (Fig. 1.3a). ACKR2 is a scavenging receptor tasked with removing inflammatory cytokines from the cell surface of LECs during inflammation [76]. This allows for suppression of immature DCs and other inflammatory cells and keeps leukocytes from adhering to LECs. ACKR4 is thought to control the distribution of CCL19 and CCL21 to assist DC migration through cognate receptor CCR7, by maintaining optimal availability of these chemokines [76].

While mouse LECs are able to adhere to plastic like FRCs, human LECs are unable to do so indicating a difference in adhesion factors or requirements that is not yet understood [77]. This might be a case of gene downregulation, as many as 50% of LEC-defining genes were found to be silenced under culture conditions compared to freshly isolated cells [78].

D'Rozario et al.

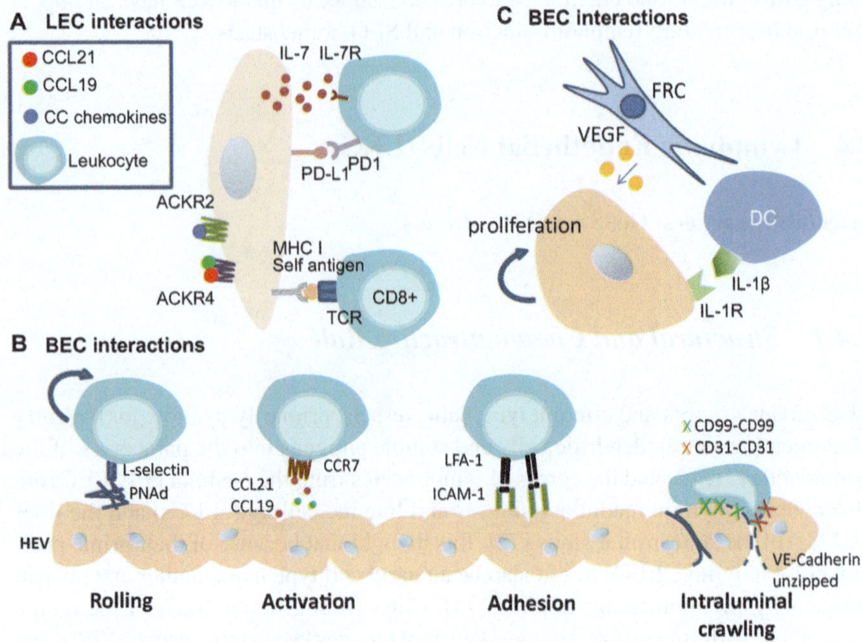

Fig. 1.3 Crosstalk between endothelial cell subsets and leukocytes. (**a**) Lymphatic endothelial cells (LECs) produce IL-7 to promote survival of naïve T cells, binding to the IL-7 receptor (IL-7R) [79]. They express programmed death ligand 1 (PD-L1) which binds PD-1 on T cells. When this interaction occurs after T cells recognise self-antigen presented by LECs via MHC class I, deletional tolerance is induced [17, 74]. LECs also control leukocyte migration and adhesion through expression of atypical cytokine receptor ACKR2, which sequesters inflammatory CC chemokines, and ACKR4, which binds CCL19 and CCL21 [107]. (**b**) High endothelial venules are constructed from blood endothelial cells (BECs) expressing peripheral node addressin (PNAd) [58]. Naïve leukocytes enter the lymph node through interactions with HEVs. First, they roll and loosely tether to the HEV when L-selectin binds PNAd on the HEV. Next, the T cell undergoes chemokine-mediated activation when CCL21 and CCL19 secreted into the lumen of the vessel bind CCR7. This induces conformational changes to LFA-1, allowing it to undergo tight adhesion by binding ICAM-1. Lastly, homeotypic interactions between CD31 and CD99, each expressed by both the leukocyte and endothelial cell, position the leukocyte between two endothelial cells, where VE-cadherin junctions unzip allowing the leukocyte to move through [84]. (**c**) BECs undergo homeostatic proliferation through signals with FRCs, which produce VEGF, and dendritic cells, through mechanisms that may include secretion of IL-1b [58, 93]

1.4.2 Interactions with T Cells and DCs

1.4.2.1 Provision of Survival Factors

LECs are a robust source of IL-7 and together with FRCs, which also produce IL-7, likely to be important regulators of T cell homeostasis [79] (Fig. 1.3a). IL-7 has proliferative and anti-apoptotic signalling abilities and is important for T cell survival in SLOs. In vitro experiments of co-cultures of LECs with T cells or T cells with conditioned media from LECs show an improved ability to promote T cell survival compared to a similar setup with the addition of anti-IL7 neutralising antibody [50].

1.4.2.2 Suppressive Tolerance

LECs are capable of suppressing the proliferation of newly activated CD4 and CD8+ T cells through the production of nitric oxide, similar to FRCs [20] (Fig. 1.2a). Under inflammatory conditions, but in the absence of infection, LECs suppress maturation of DCs, reducing expression of CD86 and their ability to prime CD8+ T cells [80]. This occurred through binding between ICAM-1 on LECs and Mac-1 on DCs and is hypothesised to reduce the risk of immune priming under inflammatory conditions in the absence of infection [80].

LECs show upregulation of surface MHC class II 18 h after initiation of an inflammatory response [22]. This may contribute to CD4 T cell tolerance through increased antigen presentation. This upregulation is IFNg dependent [26]. LECs are capable of transiently acquiring peptide-MHC class II from DCs in vivo and in vitro, directly proportional to the number of DCs present [26]. In the same study, LECs were shown to promote apoptosis of CD4+ T cells in an antigen-specific fashion [26].

Recently, Hirosue et al. [74] demonstrated that LECs can absorb exogenous OVA and will process and cross-present the OVA-derived SIINFEKL peptide fragment to CD8+ T cells in vitro [74] (Fig. 1.3a). LECs also upregulated PD-L1, which signals to PD-1 expressed by T cells and is a well-known cause of T cell exhaustion under conditions of prolonged inflammation (Fig. 1.3a). OVA-specific CD8+ T cell activation was impaired; T cells stimulated by LECs made less IL-2 and upregulated CTLA-4 earlier than those activated by DCs [74].

1.4.2.3 Deletional Tolerance

Similar to FRCs, LECs within lymph nodes express a variety of peripheral tissue-restricted antigens (PTAs), including Deaf-1 controlled Ins2 and Ppy [48, 81], though FRCs and LECs do not express identical arrays of PTAs [17, 18]. PTA expression by tolerance-inducing cells, such as LECs, is pivotal to delete T cells reactive to endogenous antigens expressed in relatively few tissues. Cohen et al. [17] showed using an endogenous melanocyte-specific self-antigen derived from tyrosinase that LECs directly presented self-antigen to Tyr-specific CD8+ T cells, deleting those that respond and purging the repertoire of autoimmune clones (Figs. 1.2b and 1.3a). In contrast, LECs indirectly induce CD4+ T cell anergy by presentation of PTAs to DCs [82], showing that LECs utilise different mechanisms of tolerance induction for CD4+ and CD8+ T cells.

In the thymus, Aire controls the expression of PTAs, but in lymph node stromal cells, Aire is not expressed [18]. However, studies in NOD mice demonstrate that PTA genes that are Aire-controlled in the thymus, such as Ambp, Fgb and Ppy, are regulated by the transcription factor Deaf-1 in the lymph node [48], which is expressed by LECs as well as FRCs [18]. During the progression of diabetes, alternative splicing of Deaf-1 occurs reducing PTA expression in mice and humans, but the effect of alternative splicing and PTA expression with respect to development of diabetes and other diseases is yet to be determined [48].

LECs do not express CD80, CD86, OX40L, 4-1BBL or CD70. These are essential molecules to drive accumulation of activated T cells, and their lack of costimulatory molecule expression may help account for their ability to delete naïve autoantigen expressing T cells. However, LECs express high levels of PD-L1, a molecule associated with deletion of tolerance-specific CD8+ T cells [17, 75].

Lymph node LECs were unique in their high expression of PTAs and PD-L1, compared with LECs in peripheral tissues such as the diaphragm and colon, showing that the lymph node microenvironment is uniquely specialised for tolerance induction [75].

1.5 Blood Endothelial Cells (BECs)

Selectable markers: Gp38-, CD31+, ICAM-1+

1.5.1 Transendothelial Migration

BECs facilitate the migration of naïve lymphocytes into lymph nodes by forming specialised postcapillary venules known as high endothelial venules (HEVs) [3]. BECs comprising HEVs show a distinctive cuboidal morphology supported by a basement membrane and have been shown to play a specialised role in allowing lymphocyte entry to SLOs through the process known as diapedesis, transendothelial migration or leukocyte extravasation (Fig. 1.3b). Migration into lymph nodes does not require inflammation, but bears similarities with migration of leukocytes to inflamed sites [83]. HEVs act as gatekeepers for lymph nodes by creating pockets holding newly arrived lymphocytes until space in the parenchyma becomes available, granting entry at a rate proportionate to egress from the lymph node [9].

During cell circulation, naïve T and B cells enter the medulla by squeezing between tightly adherent endothelial cells forming high endothelial venules (HEVs). The process of slowing down and breaching the endothelial barrier involves well-characterised interactions including leukocyte rolling, activation, adhesion and intraluminal crawling [84], which involve targeted interactions between endothelium and T cells [85] (Fig. 1.3b).

L-selectin is a primary mediator of rolling and loose attachment to HEVs. During inflammation, cytokines also upregulate expression of P-selectin and E-selectin by HEVs [3]. L-selectin binds peripheral node addressin (PNAd), referring to a group of sialomucins including CD34, podocalyxin, endomucin and nepmucin (mice and humans), as well as Glycam-1 (mice only). PNAd is expressed only by BECs comprising HEVs [58]. These glycoproteins are heavily sialylated, fucosylated and sulphated, in part through the activity of HEV-restricted GlcNAc-6-sulfotransferase [86]. Next, lymphocytes undergo integrin-mediated arrest primarily involving LFA-1 [87], binding endothelial ICAM-1 molecules, which cluster beneath the T cell to anchor it. Signalling to ICAM-1 also initiates intracellular processes that prepare endothelial cells for transendothelial migration. Next, PECAM (CD31) and CD99, each expressed by both leukocytes and endothelial cells with homophilic affinity, arrest leukocytes near the junction of adjacent endothelial cells [84]. Loss of CD31 arrests leukocytes at the junction, while loss of CD99 arrests leukocytes after they have begun to enter the junction [84]. Lastly, adherens junctions joining endothelial cells are disassembled through phosphorylation of VE-cadherin, which occurs downstream of ICAM-1 signalling and SHP2 recruitment [84] (Fig. 1.3b).

In mice, but not humans, BECs produce CCL21, which assists with arrest of rolling lymphocytes by binding G protein-coupled receptor CCR7 expressed by naïve T cells [3, 88] and may drive lymphocyte migration across the HEV barrier into the paracortex [89]. During inflammation, lymphatics bring an influx of pro-inflammatory chemokines including CCL2 and CXCL9, which are transported through conduits to the HEV lumen to increase the influx of T cells, B cells, NK cells and monocytes [3].

HEV barriers into the lymph node are thought to be regulated by atypical cytokine receptor 1 (ACKR1) that is responsible for the transport of chemokines in vivo. However, more experimental evidence is needed to substantiate this theory, as knockout ACKR1 knockout mice exhibit a varied phenotype [76].

1.5.2 BEC Proliferation and Homeostasis

Mice with endothelial cell-specific ablation of LTbR (VE-cadherin-Cre × Ltbr $^{fl/fl}$) showed reduced lymph node formation, altered HEV phenotype and a reduction of lymphocytes entering lymphatic organs [90]. Endothelial cells lost their cuboidal shape and polarisation with reduced ICAM-1 expression. An overall reduction in homing of lymphocytes was recorded, but T cell motility once inside the lymph node was not affected [90]. BECs have also been shown to increase in number during inflammation and are responsible in maintaining vascular integrity and haemostasis [34, 35, 91, 92].

Like FRCs and LECs, BECs are also important for network remodelling of the lymph node. Stromal proliferation is common for FRCs, LECs and BECs upon infection, and FRCs and HEV BECs appear to begin proliferating simultaneously as early as 2 days postinfection [58]. FRCs are the highest producers of vascular endothelial growth factor (VEGF) and are likely to contribute to growth of BECs through this molecule [58] (Fig. 1.3c).

DCs also regulate HEV phenotype and function (Fig. 1.3c). Webster et al. [93] showed that mice depleted of DCs (CD11c-DTR mice) showed a significant decrease in lymph node size and endothelial cell proliferation after injection of OVA/CFA, compared to controls. In addition, RAG1$^{-/-}$ mice that exhibit decreased cellularity and basal levels of endothelial cells still showed size increase of lymph nodes upon injection of bone marrow-derived DCs [93]. IL-1β is thought to play a partial role in in inducing endothelial cell proliferation by DCs, but not T and B cells, though other factor are yet to be identified [94]. Subsequent studies also show that T and B cells, while dispensable in initial endothelial proliferation, play a major role in subsequent maintenance and expansion of the lymph node [58].

Recent transcriptomic analysis has revealed the need for caution when using cultured BECs and LECs. The transcriptional profile of cultured BECs resembles that of LECs and might lead to misidentification during experimental procedures [78]. Specifically, more than 65% of genes selectively expressed by BECs in vivo are downregulated during culture, including MHC class II, E-selectin and ICAM-1 [78]. However, LECs maintain podoplanin and CD146 gene expression in culture, allowing differentiation from BECs [78].

1.6 Follicular Dendritic Cells (FDCs)

Selectable markers: CD35+, CD21+ FDC M1+, CXCL13+ ICAM1+ VCAM1+ BAFF+

1.6.1 Interactions with B Cells

FDCs are non-haematopoietic stromal cells contained in B cell follicles within lymphoid organs, where they are able to capture incoming antigen and store it long-term for presentation to B cells while also producing the B cell survival factors BAFF and IL-6 [42, 95–97]. FDCs play critical roles in the formation of efficient germinal centres and efficient somatic hypermutation of B cells and subsequent production of high-affinity antibodies [98, 99]. B cells can also directly capture antigen from FDCs [100], and elimination of FDCs eliminates germinal centres [98, 99, 101, 102].

FDCs are the major source of CXCL13 in lymph nodes, a chemokine which plays a crucial role in organising B follicles and the formation of germinal centres [43, 103, 104]. CXCL13 also signals to B cells and T helper cells leading them towards the follicles [43], and it enhances B cell activation [54]. FDCs retain intact antigen for extended periods (up to 12 months), which facilitates germinal centre maintenance and B cell somatic hypermutation [42].

Activated B cells migrate to the border of the follicle to present antigen to T helper cells, which provide essential costimulatory signals. B cells receiving help then migrate to the follicle's centre to proliferate and undergo hypermutation and are then subjected to selection by FDCs on the basis of recognition of antigens displayed by FDCs. Activated B cells interact with antigen presented on FDCs in a process called affinity selection, after which they progress to one of several fates (to proliferate further, class-switch or become plasma or memory B cells), while non-responding B cells become apoptotic [42].

FDCs are of enormous immunological relevance; accordingly there is an extensive literature on FDCs (see Heesters et al. [42]) that unfortunately cannot be discussed at length in this review.

1.7   Double-Negative Stromal Cells and Podoplanin-Negative Pericytes

Selectable markers: Gp38-, CD31-, ITGA7+, calponin-1

1.7.1   Identification and Characterisation

GP38-CD31 double-negative (DN) stroma represent approximately 10–20% of non-haematopoietic lymph node stromal cells [14, 18, 59]. Until recently the lineage, location and function of these cells were undescribed.

Malhotra et al. [22] used transcriptomics of sorted lymph node stromal subsets to identify a high similarity between the DN cells and fibroblastic reticular cells (FRCs), including similarities in chemokine, cytokine and growth factor expression. FRCs showed higher expression of CCL19 and CCL21 [22], though this may have been due to the heterogenous nature of the DN pool, as not all DN cells were likely to be fibroblastic in origin. A noteworthy difference was that IL-7 expression was restricted to FRCs and lacking in DN cells, while DN cells showed higher expression of the genes responsible for structure and contractile functions including higher expression of actin subtypes and myosin chain genes, which usually control smooth muscle contraction [22].

Accordingly, approx. 50% of DN cells were identified as a specialised subset of myofibroblastic pericytes, localised using specific expression of calponin-1 and integrin α7 [22]. Both calponin-1 and integrin α7 contribute to muscular function; calponin-1 is a specific actin protein that regulates force in contractile cells, especially smooth muscle cells [105], and integrin α7 connects the extracellular matrix with muscle fibres [106]. Staining with these antibodies identified this subset of double-negative cells around certain vessels in the cortex and the medulla [22]. The subset was named integrin alpha-7+ pericytes [22]. Their function is undescribed, but shared expression of many immunologically relevant molecules with FRCs may suggest similar functions. The other 50% of cells comprising the DN subset are undescribed.

1.8   Conclusions

Lymph node stromal cells form crucial roles in maintaining lymph node structure and homeostasis and have evolved to become major contributors to the immune system via their cellular interactions. This area of study has instigated a paradigm shift in the study of tolerance and has increased our understanding on immune cellular interactions. Evidence of biologically significant crosstalk occurring between stromal cells and the immune system continues to emerge, including interactions that rebuild the immune system after damage or prevent autoimmunity. These findings continue to demonstrate the importance of further research into stroma from lymphoid organs.

References

1. Buettner M, Bode U. Lymph node dissection–understanding the immunological function of lymph nodes. Clin Exp Immunol. 2012;169:205–12.
2. Bazemore AW, Smucker DR. Lymphadenopathy and malignancy. Am Fam Physician. 2002;66:2103–10.
3. Girard J-P, Moussion C, Förster R. HEVs, lymphatics and homeostatic immune cell trafficking in lymph nodes. Nat Rev Immunol. 2012;12:762–73.
4. Willard-Mack CL. Normal structure, function, and histology of lymph nodes. Toxicol Pathol. 2006;34:409–24.
5. Roozendaal R, Mebius RE. Stromal cell-immune cell interactions. Annu Rev Immunol. 2011;29:23–43.
6. Ahrendt M, Hammerschmidt SI, Pabst O, Pabst R, Bode U. Stromal cells confer lymph node-specific properties by shaping a unique microenvironment influencing local immune responses. J Immunol. 2008;181:1898–907.
7. Tomura M, et al. Tracking and quantification of dendritic cell migration and antigen trafficking between the skin and lymph nodes. Sci Rep. 2014;4:6030.
8. Bajenoff M, et al. Stromal cell networks regulate lymphocyte entry, migration and territoriality in lymph nodes. Immunity. 2006;25:989–1001.
9. Mionnet C, et al. High endothelial venules as traffic control points maintaining lymphocyte population homeostasis in lymph nodes. Blood. 2011;118:6115–22.
10. Luther SA, Tang HL, Hyman PL, Farr AG, Cyster JG. Coexpression of the chemokines ELC and SLC by T zone stromal cells and deletion of the ELC gene in the plt/plt mouse. Proc Natl Acad Sci U S A. 2000;97:12694–9.
11. Luther SA, et al. Differing activities of homeostatic chemokines CCL19, CCL21, and CXCL12 in lymphocyte and dendritic cell recruitment and lymphoid neogenesis. J Immunol. 2002;169:424–33.
12. Westermann J, et al. Naive, effector, and memory T lymphocytes efficiently scan dendritic cells in vivo: contact frequency in T cell zones of secondary lymphoid organs does

not depend on LFA-1 expression and facilitates survival of effector T cells. J Immunol. 2005;174:2517–24.

13. Lee J-W, et al. Peripheral antigen display by lymph node stroma promotes T cell tolerance to intestinal self. Nat Immunol. 2007;8:181–90.

14. Link A, et al. Fibroblastic reticular cells in lymph nodes regulate the homeostasis of naive T cells. Nat Immunol. 2007;8:1255–65.

15. Magnusson FC, et al. Direct presentation of antigen by lymph node stromal cells protects against CD8 T-cell-mediated intestinal autoimmunity. Gastroenterology. 2008;134:1028–37.

16. Molenaar R, et al. Lymph node stromal cells support dendritic cell-induced gut-homing of T cells. J Immunol. 2009;183:6395–402.

17. Cohen JN, et al. Lymph node-resident lymphatic endothelial cells mediate peripheral tolerance via Aire-independent direct antigen presentation. J Exp Med. 2010;207:681–8.

18. Fletcher AL, et al. Lymph node fibroblastic reticular cells directly present peripheral tissue antigen under steady-state and inflammatory conditions. J Exp Med. 2010;207:689–97.

19. Khan O, et al. Regulation of T cell priming by lymphoid stroma. PLoS One. 2011;6:e26138.

20. Lukacs-Kornek V, et al. Regulated release of nitric oxide by nonhematopoietic stroma controls expansion of the activated T cell pool in lymph nodes. Nat Immunol. 2011;12:1096–104.

21. Siegert S, et al. Fibroblastic reticular cells from lymph nodes attenuate T cell expansion by producing nitric oxide. PLoS One. 2011;6:e27618.

22. Malhotra D, et al. Transcriptional profiling of stroma from inflamed and resting lymph nodes defines immunological hallmarks. Nat Immunol. 2012;13:499–510.

23. Baptista AP, et al. Lymph node stromal cells constrain immunity via MHC class II self-antigen presentation. Elife. 2014;3:e04433.

24. Cremasco V, et al. B cell homeostasis and follicle confines are governed by fibroblastic reticular cells. Nat Immunol. 2014;15:973–81.

25. Denton AE, Roberts EW, Linterman MA, Fearon DT. Fibroblastic reticular cells of the lymph node are required for retention of resting but not activated CD8+ T cells. Proc Natl Acad Sci U S A. 2014;111:12139–44.

26. Dubrot J, et al. Lymph node stromal cells acquire peptide-MHCII complexes from dendritic cells and induce antigen-specific CD4+ T cell tolerance. J Exp Med. 2014;211:1153–66.

27. Fletcher AL, Acton SE, Knoblich K. Lymph node fibroblastic reticular cells in health and disease. Nat Rev Immunol. 2015;15:350–61.

28. Acton SE, et al. Dendritic cells control fibroblastic reticular network tension and lymph node expansion. Nature. 2014;514:498–502.

29. Astarita JL, et al. The CLEC-2-podoplanin axis controls the contractility of fibroblastic reticular cells and lymph node microarchitecture. Nat Immunol. 2015;16:75–84.

30. Roozendaal R, et al. Conduits mediate transport of low-molecular-weight antigen to lymph node follicles. Immunity. 2009;30:264–76.

31. Lammermann T, et al. Rapid leukocyte migration by integrin-independent flowing and squeezing. Nature. 2008;453:51–5.

32. Sixt M, et al. The conduit system transports soluble antigens from the afferent lymph to resident dendritic cells in the T cell area of the lymph node. Immunity. 2005;22:19–29.

33. Gretz JE, Norbury CC, Anderson AO, Proudfoot AEI, Shaw S. Lymph-borne chemokines and other low molecular weight molecules reach high endothelial venules via specialized conduits while a functional barrier limits access to the lymphocyte microenvironments in lymph node cortex. J Exp Med. 2000;192:1425–40.

34. Herzog BH, et al. Podoplanin maintains high endothelial venule integrity by interacting with platelet CLEC-2. Nature. 2013;502:105–9.

35. Yang C-Y, et al. Trapping of naive lymphocytes triggers rapid growth and remodeling of the fibroblast network in reactive murine lymph nodes. Proc Natl Acad Sci U S A. 2014;111:E109–18.

36. Katakai T, Hara T, Sugai M, Gonda H, Shimizu A. Lymph node fibroblastic reticular cells construct the stromal reticulum via contact with lymphocytes. J Exp Med. 2004;200:783–95.

37. Mionnet C, et al. Identification of a new stromal cell type involved in the regulation of inflamed B cell follicles. PLoS Biol. 2013;11:e1001672.
38. Fasnacht N, et al. Specific fibroblastic niches in secondary lymphoid organs orchestrate distinct Notch-regulated immune responses. J Exp Med. 2014;211:2265–79.
39. Katakai T, et al. Organizer-like reticular stromal cell layer common to adult secondary lymphoid organs. J Immunol. 2008;181:6189–200.
40. Katakai T. Marginal reticular cells: a stromal subset directly descended from the lymphoid tissue organizer. Front Immunol. 2012;3:200.
41. Jarjour M, et al. Fate mapping reveals origin and dynamics of lymph node follicular dendritic cells. J Exp Med. 2014;211:1109–22.
42. Heesters B a, Myers RC, Carroll MC. Follicular dendritic cells: dynamic antigen libraries. Nat Rev Immunol. 2014;14:495–504.
43. Ansel KM, et al. A chemokine-driven positive feedback loop organizes lymphoid follicles. Nature. 2000;406:309–14.
44. Asperti-Boursin F, Real E, Bismuth G, Trautmann A, Donnadieu E. CCR7 ligands control basal T cell motility within lymph node slices in a phosphoinositide 3-kinase- independent manner. J Exp Med. 2007;204:1167–79.
45. Bogdan C. Nitric oxide and the immune response. Nat Immunol. 2001;2:907–16.
46. Reynoso ED, et al. Intestinal tolerance is converted to autoimmune enteritis upon PD-1 ligand blockade. J Immunol. 2009;182:2102–12.
47. Anderson MS, et al. Projection of an immunological self shadow within the thymus by the aire protein. Science. 2002;298:1395–401.
48. Yip L, et al. Deaf1 isoforms control the expression of genes encoding peripheral tissue antigens in the pancreatic lymph nodes during type 1 diabetes. Nat Immunol. 2009;10:1026–33.
49. Chai Q, et al. Maturation of lymph node fibroblastic reticular cells from myofibroblastic precursors is critical for antiviral immunity. Immunity. 2013;38:1013–24.
50. Onder L, et al. IL-7-producing stromal cells are critical for lymph node remodeling. Blood. 2012;120:4675–83.
51. Zeng M, et al. Cumulative mechanisms of lymphoid tissue fibrosis and T cell depletion in HIV-1 and SIV infections. J Clin Invest. 2011;121:998–1008.
52. Zeng M, et al. Critical role of CD4 T cells in maintaining lymphoid tissue structure for immune cell homeostasis and reconstitution. Blood. 2012;120:1856–67.
53. Mackay F, Schneider P. Cracking the BAFF code. Nat Rev Immunol. 2009;9:491–502.
54. Sáez de Guinoa J, Barrio L, Mellado M, Carrasco YR. CXCL13/CXCR5 signaling enhances BCR-triggered B-cell activation by shaping cell dynamics. Blood. 2011;118:1560–9.
55. Suenaga F, et al. Loss of lymph node fibroblastic reticular cells and high endothelial cells is associated with humoral immunodeficiency in mouse graft-versus-host disease. J Immunol. 2014. https://doi.org/10.4049/jimmunol.1401022.
56. Acton SE, et al. Podoplanin-rich stromal networks induce dendritic cell motility via activation of the C-type lectin receptor CLEC-2. Immunity. 2012;37:276–89.
57. Kumar V, et al. A dendritic-cell-stromal axis maintains immune responses in lymph nodes. Immunity. 2015;42:719–30.
58. Chyou S, et al. Coordinated regulation of lymph node vascular-stromal growth first by CD11c+ cells and then by T and B cells. J Immunol. 2011;187:5558–67.
59. Fletcher AL, et al. Reproducible isolation of lymph node stromal cells reveals site-dependent differences in fibroblastic reticular cells. Front Immunol. 2011;2:35.
60. Dougall WC, et al. RANK is essential for osteoclast and lymph node development. Genes Dev. 1999;13:2412–24.
61. Kong YY, et al. OPGL is a key regulator of osteoclastogenesis, lymphocyte development and lymph-node organogenesis. Nature. 1999;397:315–23.
62. Knoop KA, Butler BR, Kumar N, Newberry RD, Williams IR. Distinct developmental requirements for isolated lymphoid follicle formation in the small and large intestine: RANKL is essential only in the small intestine. Am J Pathol. 2011;179:1861–71.

63. Benezech C, et al. Ontogeny of stromal organizer cells during lymph node development. J Immunol. 2010;184:4521–30.
64. Cupedo T, et al. Presumptive lymph node organizers are differentially represented in developing mesenteric and peripheral nodes. J Immunol. 2004;173:2968–75.
65. Luther SA, Ansel KM, Cyster JG. Overlapping roles of CXCL13, interleukin 7 receptor alpha, and CCR7 ligands in lymph node development. J Exp Med. 2003;197:1191–8.
66. Ohl L, et al. Cooperating mechanisms of CXCR5 and CCR7 in development and organization of secondary lymphoid organs. J Exp Med. 2003;197:1199–204.
67. Eberl G, et al. An essential function for the nuclear receptor RORgamma(t) in the generation of fetal lymphoid tissue inducer cells. Nat Immunol. 2004;5:64–73.
68. Katakai T, et al. A novel reticular stromal structure in lymph node cortex: an immuno-platform for interactions among dendritic cells, T cells and B cells. Int Immunol. 2004;16:1133–42.
69. Krautler NJ, et al. Follicular dendritic cells emerge from ubiquitous perivascular precursors. Cell. 2012;150:194–206.
70. Scandella E, et al. Restoration of lymphoid organ integrity through the interaction of lymphoid tissue-inducer cells with stroma of the T cell zone. Nat Immunol. 2008;9:667–75.
71. Braun A, et al. Afferent lymph-derived T cells and DCs use different chemokine receptor CCR7-dependent routes for entry into the lymph node and intranodal migration. Nat Immunol. 2011;12:879–87.
72. Ulvmar MH, et al. The atypical chemokine receptor CCRL1 shapes functional CCL21 gradients in lymph nodes. Nat Immunol. 2014;15:623–30.
73. Pham THM, et al. Lymphatic endothelial cell sphingosine kinase activity is required for lymphocyte egress and lymphatic patterning. J Exp Med. 2010;207:17–27.
74. Hirosue S, et al. Steady-state antigen scavenging, cross-presentation, and CD8+ T cell priming: a new role for lymphatic endothelial cells. J Immunol. 2014;192(11):5002.
75. Cohen JN, et al. Tolerogenic properties of lymphatic endothelial cells are controlled by the lymph node microenvironment. PLoS One. 2014;9:e87740.
76. Nibbs RJB, Graham GJ. Immune regulation by atypical chemokine receptors. Nat Rev Immunol. 2013;13:815–29.
77. Fletcher AL, et al. Reproducible isolation of lymph node stromal cells reveals site-dependent differences in fibroblastic reticular cells. Front Immunol. 2011;2:35.
78. Amatschek S, et al. Blood and lymphatic endothelial cell-specific differentiation programs are stringently controlled by the tissue environment. Blood. 2007;109:4777–85.
79. Miller CN, et al. IL-7 production in murine lymphatic endothelial cells and induction in the setting of peripheral lymphopenia. Int Immunol. 2013;25:471–83.
80. Podgrabinska S, et al. Inflamed lymphatic endothelium suppresses dendritic cell maturation and function via Mac-1/ICAM-1-dependent mechanism. J Immunol. 2009;183:1767–79.
81. Rouhani SJ, Eccles JD, Tewalt EF, Engelhard VH. Regulation of T-cell tolerance by lymphatic endothelial cells. J Clin Cell Immunol. 2014;5242.
82. Rouhani SJ, et al. Roles of lymphatic endothelial cells expressing peripheral tissue antigens in CD4 T-cell tolerance induction. Nat Commun. 2015;6:6771.
83. Faveeuw C, Preece G, Ager A. Transendothelial migration of lymphocytes across high endothelial venules into lymph nodes is affected by metalloproteinases. Blood. 2001;98:688–95.
84. Muller WA. The regulation of transendothelial migration: new knowledge and new questions. Cardiovasc Res. 2015;107:310–20.
85. Sage PT, Carman CV. Settings and mechanisms for trans-cellular diapedesis. Front Biosci (Landmark Ed). 2009;14:5066–83.
86. Bistrup A, et al. Sulfotransferases of two specificities function in the reconstitution of high endothelial cell ligands for L-selectin. J Cell Biol. 1999;145:899–910.
87. Warnock RA, Askari S, Butcher EC, von Andrian UH. Molecular mechanisms of lymphocyte homing to peripheral lymph nodes. J Exp Med. 1998;187:205–16.
88. Stein JV, et al. The CC chemokine thymus-derived chemotactic agent 4 (TCA-4, secondary lymphoid tissue chemokine, 6Ckine, exodus-2) triggers lymphocyte function-associated antigen 1-mediated arrest of rolling T lymphocytes in peripheral lymph node high endothelial venules. J Exp Med. 2000;191:61–76.

89. Förster R, et al. CCR7 coordinates the primary immune response by establishing functional microenvironments in secondary lymphoid organs. Cell. 1999;99:23–33.
90. Onder L, et al. Endothelial cell-specific lymphotoxin-β receptor signaling is critical for lymph node and high endothelial venule formation. J Exp Med. 2013;210:465–73.
91. Lee M, et al. Transcriptional programs of lymphoid tissue capillary and high endothelium reveal control mechanisms for lymphocyte homing. Nat Immunol. 2014;15:982–95.
92. Chyou S, et al. Fibroblast-type reticular stromal cells regulate the lymph node vasculature. J Immunol. 2008;181:3887–96.
93. Webster B, et al. Regulation of lymph node vascular growth by dendritic cells. J Exp Med. 2006;203:1903–13.
94. Benahmed F, et al. Multiple CD11c+ cells collaboratively express IL-1β to modulate stromal vascular endothelial growth factor and lymph node vascular-stromal growth. J Immunol. 2014;192:4153–63.
95. Bajénoff M, Germain RN. B-cell follicle development remodels the conduit system and allows soluble antigen delivery to follicular dendritic cells. Blood. 2009;114:4989–97.
96. Wu Y, et al. IL-6 produced by immune complex-activated follicular dendritic cells promotes germinal center reactions, IgG responses and somatic hypermutation. Int Immunol. 2009;21:745–56.
97. Wang X, et al. Follicular dendritic cells help establish follicle identity and promote B cell retention in germinal centers. J Exp Med. 2011;208:2497–510.
98. Matsumoto M, et al. Affinity maturation without germinal centres in lymphotoxin-alpha-deficient mice. Nature. 1996;382:462–6.
99. Wang X, et al. Follicular dendritic cells help establish follicle identity and promote B cell retention in germinal centers. J Exp Med. 2011;208:2497–510.
100. Suzuki K, Grigorova I, Phan TG, Kelly LM, Cyster JG. Visualizing B cell capture of cognate antigen from follicular dendritic cells. J Exp Med. 2009;206:1485–93.
101. Fischer MB, et al. Dependence of germinal center B cells on expression of CD21/CD35 for survival. Science. 1998;280:582–5.
102. Gommerman JL, et al. Manipulation of lymphoid microenvironments in nonhuman primates by an inhibitor of the lymphotoxin pathway. J Clin Invest. 2002;110:1359–69.
103. Gunn MD, et al. A B-cell-homing chemokine made in lymphoid follicles activates Burkitt's lymphoma receptor-1. Nature. 1998;391:799–803.
104. Legler DF, et al. B cell-attracting chemokine 1, a human CXC chemokine expressed in lymphoid tissues, selectively attracts B lymphocytes via BLR1/CXCR5. J Exp Med. 1998;187:655–60.
105. Yamamura H, Hirano N, Koyama H, Nishizawa Y, Takahashi K. Loss of smooth muscle calponin results in impaired blood vessel maturation in the tumor? Host microenvironment. Cancer Sci. 2007;98:757–63.
106. Mayer U, et al. Absence of integrin alpha 7 causes a novel form of muscular dystrophy. Nat Genet. 1997;17:318–23.
107. Nibbs RJB, Graham GJ. Immune regulation by atypical chemokine receptors. Nat Rev Immunol. 2013;13:815–29.

Chapter 2
Stromal Cell Responses in Infection

Paul M. Kaye

Abstract Stromal cells and the immune functions that they regulate underpin multiple aspects of host defence, but the study of stromal cells as targets of infection and as regulators of anti-infective immunity is in its infancy and still limited to a few well-worked examples. In this review, the role of stromal cells at each sequential stage of infection is discussed, with examples drawn from across the spectrum of infectious agents, from prions to the parasitic helminths. Gaps in knowledge are identified, the challenges in studying stromal cell biology in the context of infection are highlighted, and the potential for stromal cell-targeted therapeutics is briefly discussed.

Keywords Stromal infection · Innate immunity · TLOs · Stromal architecture · Stromal APCs · Inflammation resolution

2.1 Introduction

The pathogenesis of infectious disease is complex and involves a myriad of processes, some but not all related to core immune mechanisms, most of which in one form or another are underpinned by features of stromal cell biology. Stromal cells provide the tissue architecture at the primary interface with infectious agents (e.g. the skin or mucosa), act as a potential cellular target for infection, provide the framework for the compartmentalized functions of lymphoid tissue and immune response induction and generate and maintain the vascular environment that allows for effector cell trafficking to the sites of infection. At the end of infection, stromal cells play a role in resolution of immune-mediated pathology and the return to homoeostasis. Details of many of these functions of stromal cells during development, under homoeostatic conditions and during cancer are described in

P. M. Kaye (✉)
Centre for Immunology and Infection, Department of Biology and Hull York Medical School, University of York, York, UK
e-mail: paul.kaye@york.ac.uk

© Springer International Publishing AG, part of Springer Nature 2018
B. M.J Owens, M. A. Lakins (eds.), *Stromal Immunology*, Advances in Experimental Medicine and Biology 1060, https://doi.org/10.1007/978-3-319-78127-3_2

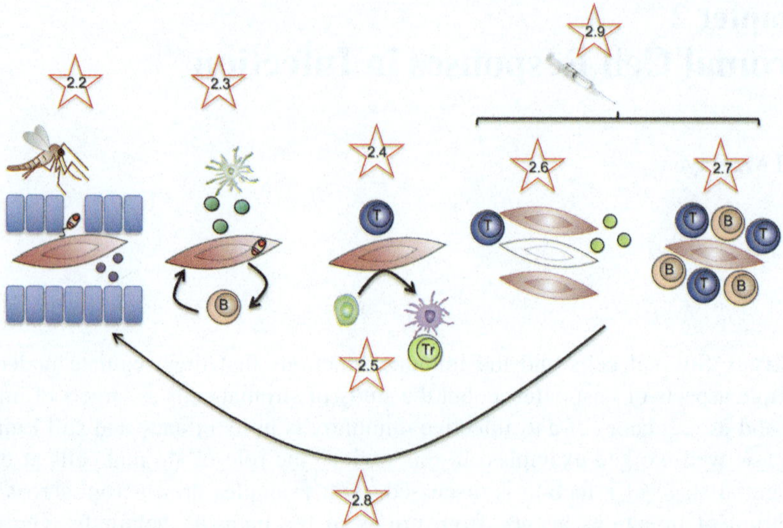

Fig. 2.1 The temporal role of stromal cells in regulating immunity to infection. Schematic shows a stylized time course depicting the events from initial pathogen contact through to disease resolution that may be influenced by stromal cell interactions and functions. For examples of stromal cell interactions with specific pathogens at each stage (numbered stars), see main text

detail elsewhere in this volume. Here, the focus will be on providing key exemplars of how stromal cells interact, directly or indirectly, with pathogens and help to orchestrate subsequent inflammatory and immune responses that ultimately lead either to pathogen elimination or the establishment of a chronic persistent infection. The broader role of stromal cells in the resolution of infection-associated pathology is also discussed, although to date there are few studies addressing this important aspect of infection biology. For the sake of brevity, discussion regarding stromal cell interactions with pathogens associated with the development of cancer (e.g. Epstein-Barr Virus) has been omitted (Fig. 2.1).

2.2 Initiation of Infection: Stromal Cells as Targets for Adhesion and Infection

For successful infection to begin, pathogens require means to adhere to and/or penetrate external barriers. For mucosal pathogens, epithelial cells represent a major site of pathogen attachment at mucosal surfaces, and an intact epithelium provides a barrier to direct interaction between these infectious agents and underlying mesenchymal stromal cells. The role of crosstalk between epithelial cells and stromal cells, serving as an integral part of the signaling required for epithelial barrier function and maintenance, is well documented, notably in the female reproductive tract and mammary gland [1, 2]. This establishes a paradigm likely to be operating at most epithelial sites and suggests that any disruption to this functional unit may occur through either epithelial or stromal cell changes.

Some mucosal pathogens such as *Entamoeba histolytica* [3] are professional tissue invaders and use a variety of molecular and cellular strategies to penetrate deep into the mucosa and submucosa; yet little attention has been paid to the consequences of such local tissue trauma for stromal cell function. Gastrointestinal worms have a variety of ways to interact with and manipulate the mucosal epithelium and otherwise disrupt local immunity [4, 5]; however, the direct action of helminth-derived molecules on intestinal stromal cells has not been reported in any detail.

Many of the major pathogens of man are transmitted via breaks in skin barrier that occur during blood feeding by their arthropod vector. For these pathogens, there may be more direct and immediate access to tissue stromal cells, and the tropism of intracellular pathogens to stromal cells after epithelial barrier breach provides an opportunity for their establishment and long-term survival. In addition to facilitating access to stromal cells, the tissue trauma caused by the bite of haematophagous insects may also directly trigger stromal cell release of alarmins (see below) [6]. In the case of the intracellular parasites, the reduced microbicidal capacity of stromal cells compared to myeloid cells may provide a driver for this behaviour. For example, the parasitic protozoan *Trypanosoma cruzi* invades the skin and oral mucosa in a process of contaminative transmission, with infectious metacyclic parasites being deposited in the faeces of feeding triatomine bugs. *T. cruzi* has a broad host cell range, making use of a plethora of attachment molecules and active processes to invade stromal cells in a process of triggered phagocytosis [7]. Other vector-borne parasites such as *Leishmania*, whilst historically regarded as having a more limited host cell range, have also been noted as intracellular parasites of stromal cells at chronic stages of infection [8, 9]. It is not clear whether stromal cell infection also occurs early after infection and has hitherto gone unrecognized.

In experimental models of infection, stromal cell tropism is also noted and may illustrate how tropism may be both cell- and tissue-specific in nature. For example, after intraperitoneal murine cytomegalovirus (MCMV) infection, ERTR7$^+$ marginal zone reticular cells represent the major target for infection within the spleen, with subsequent viral spread to red pulp fibroblasts. In contrast, in the lymph node primary infection occurs within CD169$^+$ subcapsular sinus macrophages [10].

These various studies also raise an important question regarding the temporal regulation of adhesion and/or pathogen selective receptors on stromal cells (e.g. by mediators involved in inflammation) and whether this may also contribute to the patterns of cellular tropism that are observed. For example, in the case of corneal infection with HSV, the major viral receptor nectin-1 is initially absent from stromal cells, but expression is induced early during inflammation allowing increased viral host cell range [11, 12]. Further studies in a variety of infection models to address the role of direct stromal cell infection at different stages of the infection process would clearly help in delineating the importance of such interactions to disease progression and pathogen life cycle maintenance.

HIV infection provides an interesting and relevant example of how stromal cells can both be target for infection and regulator of infection in other cells. HIV infects follicular dendritic cells (FDC), which provide a reservoir for viral infection of CD4+ T cells and macrophages [13–15], but in addition, these infected FDCs provide TNF-dependent augmentation of HIV transcription and viral replication in CD4+ T cells [16].

Finally, to fully define pathogen cellular tropism, it may be necessary to consider the potential for lineage transformation. At least one intracellular pathogen has been shown to induce epithelial-to-mesenchymal transition (EMT). The *cag* pathogenicity island of *Helicobacter pylori* encodes a type 4 secretion system that delivers bacterial effectors into the cytosol of gastric epithelial cells and induces EMT [17]. Whether this can be induced by other intracellular pathogens remains to be determined. Another instance where lineage boundaries become blurred is the case of the fibrocyte, which shares markers of haematopoietic cells and fibroblasts [18]. Recently, the capacity of human and murine blood fibrocytes to support internalization of promastigotes of *Leishmania amazonensis* was reported, along with the capacity of these cells to support intracellular transformation to amastigotes [19]. Strikingly, fibrocytes produced high levels of NO and cleared parasites within a few days of infection, suggesting that transient waves of infection within fibrocytes in vivo could go unnoticed. Furthermore, a variety of *Leishmania* species have been shown to infect adipose-derived mesenchymal stem cells in vitro [20], though whether this alters stem cell function and differentiation capacity has yet to be determined. It seems likely that at least some intracellular pathogens will be found to have the capacity to affect the pluripotency of stem cells, but this remains an area to be explored experimentally.

2.3 Early Inflammation: Stromal Cells as Contributors to Innate Immunity

Cooperation between haematopoietic and stromal cells can play an important role in initiating inflammation. For example, TLR4 expression on stromal cells is required for optimal resistance against uropathogenic *E. coli* but is not sufficient for induction of inflammation in the absence of TLR4 expression on haematopoietic cells [21]. Protection in a lethal model of vesicular stomatitis virus (VSV) infection required intact TLR and retinoic acid-inducible gene I-like helicase (RLH) signaling in both radioresistant stromal cells and haematopoietic cells [22]. Likewise, effective immunity in a model of oropharyngeal candidiasis required expression of the NLRC4 inflammasome in radioresistant stromal cells, working cooperatively with NLRP3 inflammasome [23].

Recent studies in the well-established MCMV infection model have also provided insight into how stromal cells can play a direct role in innate immunity. Type I interferons are required for the early control of MCMV replication in the mouse spleen, and various studies have indicated that both plasmacytoid and conventional dendritic cells (DCs) play temporally discrete roles in producing the bulk of type I interferon detectable between 36 and 48 h postinfection. However, this wave of type

I interferon is preceded (2 h postinfection) by a burst of type I interferon derived from LTβR-expressing splenic marginal zone reticular cells, acting in concert with LTαβ-producing B cells [24, 25].

Alarmins represent a constitutively available group of evolutionarily diverse molecules that normally have important intracellular roles but which can be released into the extracellular environment through tissue injury or inflammatory signaling cellular roles during regulation of innate immunity and may be triggered directly by tissue damage or indirectly through the production of other early mediators of inflammation such as IL-17. Alarmins include IL-33, various S100 molecules, HMGB1 and thymic stromal lymphopoietin (TSLP) [26, 27]. The contribution of stromal cells to the production and/or regulation of alarmins in the context of epithelial cell injury during infection is, however, poorly understood.

Some of the aforementioned papers, however, highlight a conspicuous difficulty in the field of stromal cell biology: assignment of the term "stromal" to cells that are radioresistant in radiation bone marrow chimeras. Whilst this experimental approach can identify, but not distinguish between, properties attributable to radioresistant mesenchyme-derived stromal cells and epithelial cells, the recent identification of radioresistant resident tissue macrophages of yolk sac origin [28–30] introduces some significant question marks over previous attribution of cell function. The ability to generate mouse strains for lineage tracing and for selective stromal cell-targeted gene ablation (e.g. [31]) is an important step towards clarifying the function of stromal cells throughout the infection process.

2.4 Induction of Acquired Immunity: Stromal Cells as APC During Infectious Disease

The notion that antigen presentation within lymphoid tissues is restricted to haematopoietic cells has been over-turned by a number of recent studies that indicate that stromal cells have all the machinery necessary for both MHCI [32, 33] and MHCII-dependent antigen presentation [34] and in addition can acquire and functionally express MHCII-peptide complexes derived from DCs [35]. Under homoeostatic conditions, this imparts an ability to effect CD8+ T cell deletion to self-antigens and regulate the extent of CD4+ T cell priming, either directly by inducing anergy [35] or indirectly by maintaining the pool of CD4+ Tregs [34]. The extent to which pathogen-derived antigens are presented in this way by stromal cells within the lymphoid tissue microenvironment is as yet unknown but clearly warrants further investigation given the potential for this route of antigen presentation to modify the quantity (and perhaps quality) of the response to infection. Whether stromal cells outside lymphoid tissue also are endowed with these properties will be important to address, as will be the question of whether stromal cell antigen presenting function "matures" through infection, in an equivalent manner to that seen in the haematopoietic lineage. In this context, it is interesting to note that inflammation, including that driven by infection, can lead to local tissue fibroblasts recapitulating the ontogeny of lymphoid tissue fibroblasts, expressing the canonical lymph node stromal marker podoplanin (gp38) [36]. Whether these cells acquire all functions associated with their lymphoid tissue resident counterparts remains to be determined, and illustrates an experimental setting where transcriptomic analysis of stromal cells might be particularly helpful.

2.5 Maintaining the Balance: Stromal Cells and Immune Regulation

As indicated above, new evidence indicates that lymph node stromal cells can directly engage with T cells via MHCI and MHCII restricted antigen presentation, and that the primary purpose of these interactions under homeostatic conditions appears to be the induction of one or other mechanism of self-tolerance. To date, however, most attention has been focused on how stromal cells induce a regulatory environment and thus influence T cell activation indirectly rather than directly. Early studies in vitro demonstrated the capacity of stromal cell lines to drive HSPC into a programme of myelopoiesis, often generating novel subsets of dendritic like cells [37–39]. In a model of *Leishmania donovani* infection, it was shown that ex vivo isolated stromal cells were able to induce lin⁻c-kit⁺ progenitors to differenti-ate to a greater extent that stromal cells isolated from uninfected mice. Furthermore, the resulting CD11c^{lo} CD45RB$^+$ IL-10-producing DCs had potent regulatory prop-erties (defined in vitro by suppression of T cell proliferation to antigen presented by conventional CD11c DC and in vivo by the induction of antigen-specific tolerance) [9]. Subsequently, it was demonstrated that infection-associated inflammation enhanced the function of this splenic red pulp stromal haematopoietic niche, princi-pally through aberrant expression of CCL8 [40].

A challenge for studying stromal cell biology in an infection model where stro-mal cells themselves can be infected, albeit at variable frequency, is to distinguish whether any changes in stromal cell function are directly attributable to intracellular parasitism (e.g. mediated via parasite-induced host cell intrinsic changes in signal-ing pathways) or whether they reflect the action of cytokines and or other factors operating in trans in a complex inflammatory "soup". Importantly in the latter study [40], stromal cell lines were used to show that CCL8 expression was directly induced in stromal cell lines in vitro by infection with *L. donovani* and that in vivo, CCL8$^+$ ERTR-7$^+$ stromal cells also contained parasites. These data do not however rule out trans-acting factors as contributors to the in vivo response. Indeed, as endo-thelial cells and fibroblasts isolated from various tissues, with or without inflamma-tory stress, have been shown to induce regulatory myeloid cells capable of inhibiting T cell responses [41–45] and in one case to block virus-mediated activation of pDC [46], it is likely haematopoietic support is a generic tunable property of stromal cells that helps maintain immune balance.

Stromal cells have been less well studied in the context of helminth infections and immune regulation. B cells have been shown to play a role in immune regulation in some chronic infections. In the case of *Schistosoma mansoni* infection, regulatory B cell development has been linked to the production of BAFF, a cytokine produced by DC and stromal cells in response to helminth antigens [47]. Surprisingly, stromal cell modifications have also been implicated in the onward transmission of this parasite. *S. mansoni* egg excretion is essential for completion of the life cycle, and this is facilitated by entry of eggs into Peyer's patches, which respond with extensive remodeling of their stromal elements [48].

In summary, whilst it is tempting to "hand-wave" by saying that stromal cells are almost certain to have a role in the overall generation of the changing immune environment during infection, through participation in the control of lymphocyte and dendritic cell functions or through their contribution to the cytokine environment, only carefully designed and executed studies using stromal cell specific targeting of key immune mediators will provide the answer to the question of how important stromal cell responses are relative to those of other cells in driving the ultimate phenotype – pathogen elimination or persistence of infection.

2.6 Perpetuating Chronic Infection: Breakdown of Stromal Cell Architecture

As detailed elsewhere within this volume, stromal cells play a central role in the development and maintenance of lymphoid tissue architecture. On the assumption that immune architecture is therefore integral to the efficiency of the immune system, it is perhaps not surprising that many studies of disease, including infectious disease, have noted changes in lymphoid tissue architecture and associated these with dysregulation of immunity.

Examination of the lymph nodes of patients with progressive HIV infection demonstrated marked degenerative changes to germinal centres, including the depletion of the FDC network, a process termed follicle lysis [49, 50]. Of note, a study in SIV-infected macaques demonstrated that although FDCs were also greatly reduced in number in this model infection, residual FDCs appeared to make more functionally productive interactions with B cells [51]. This study serves as a reminder that pathology-associated loss of cell number should not be equated directly with loss of function. Similar structural changes to the FDC network have also been observed in non-viral infection models, including chronic visceral leishmaniasis. Here, FDC loss was determined both by immunohistochemistry and by lack of immune complex trapping within GCs [52]. Strikingly, heavily parasitized macrophages became abundant within these GCs, resembling the tingible body macrophages described in HIV infection [49].

The demonstration that a distinct population of podoplanin+ fibroblastic reticular cells was present within the T cell zone of lymphoid tissues focused attention on how this stromal cell subset was altered in a variety of infection models. In chronic HIV infection, fibrosis of lymphoid tissue becomes apparent, and this results in loss of integrity of the FRC network. Importantly, the extent of loss of FRCs and collagen deposition are predictors of the ability of highly active antiretroviral therapy (HAART) to restore T cell count. Furthermore, HAART is most effective at restoring FRC networks when given early during disease [53, 54].

FRCs are also lost during infection with *Leishmania donovani* in mice [55]. As expected, FRC deficiency was also characterized by a loss of constitutive CCL21 and CCL19 production in the spleen of infected mice and by alterations in DC and T cell traffic. Although DC migration from the marginal zone into the T cell zone was impaired in infected mice, this was not the direct result of loss of FRCs. By adoptive transfer, it was shown that the residual FRCs and CCL21-expressing endothelium were sufficient to allow migration of DC isolated from naïve mice, and whereas conversely DCs from infected mice failed to migrate even in naïve hosts. Hence in this case, loss of CCR7 expression by DCs in infected mice, which was in turn regulated in a TNF and IL-10-dependent manner, appeared to play a more dominant role in affecting DC-T cell interactions than the loss of stromal architecture per se. Of interest in this regard, computational models have been developed to assess the impact of changes in FRC network density and inter-connectivity in regulating cellular encounters between T cells and DC [56].

Unlike the situation in chronic HIV and visceral leishmaniasis, experimental viral infections provide examples of situations where extreme but transient pathology occurs. In LCMV infection, splenomegaly is transient, peaking during the first week of infection but subsiding to normal range within 10 days. Loss of FRCs accompanies splenomegaly, as does inability to respond to challenge with exogenous antigens, but the FRC network recovers remarkably quickly as infection is cleared. Restoration of this architecture is dependent upon the presence of RORγ⁺ lymphoid tissue inducer cells and LTαβ signaling, suggesting that "repair" of lymphoid tissue architecture recapitulates processes that occur during lymphoid tissue development [57]. During MCMV, disruption of lymphoid stroma appears restricted to the FRCs, which show similar alterations in gp38 staining pattern and changes in CCL21 expression as observed in visceral leishmaniasis and LCMV infection. B zone stromal cells, however, appear not to be significantly affected during this infection, as judged by maintenance of CXCL13 [58].

In addition to stromal cell changes, other microarchitectural changes often accompany splenomegaly, notably loss or displacement of "stromal" macrophages within the marginal zone. Mice infected with *L. donovani* [59], *Plasmodium chabaudi* [60] and MCMV [58] all show loss of SIGNR1⁺ marginal zone and CD169⁺ marginal metallophils to a greater of lesser extent. Collectively, these data illustrate that there is a degree of commonality in the structural changes seen irrespective of infectious agent. However, the process of remodeling may be driven by quite independent mechanisms. For example, in LCMV infection, antiviral CD8⁺ T cells destroy FRCs [57]; in *P. chabaudi* infection, CD8⁺ T cells selectively kill CD169⁺ marginal metallophils through a perforin and Fas-dependent pathway [60]; and during *L. donovani* infection, SIGNR1⁺ marginal zone macrophages are lost in a TNF-dependent manner [59]. The precise impact that can be attributed to changes in secondary lymphoid tissue, in terms of immunocompetence, may not be possible to discern from simple single infection models such as those described above and may require adaptation of various co-infection models to become fully apparent.

The chronicity and extent of *Leishmania donovani*-induced splenomegaly also provide a context in which pathologic angiogenesis occurs to an exaggerated extent. Vascular remodeling in this disease requires the coordinated action of distinct myeloid cell populations, working in a compartment-specific manner. Thus, inflammatory monocytes regulated the expansion of the red pulp vasculature [61], whereas a population of "resident" macrophages, which were found bordering the denuded marginal zone, play a role in inducing neo-angiogenesis in the white pulp. Of note, this localized angiogenesis is controlled by the aberrant expression of the neurotrophin Bndf and its receptor Ntrk2 (Trkb) on macrophages and endothelial cells, respectively [62].

2.7 Perpetuating the Response: Ectopic Lymphoid Structures

Chronic inflammation is often associated with the generation of ectopic or tertiary lymphoid tissue, a topic summarized elsewhere in this volume. Not surprisingly, therefore, infections may also give rise to these structures, although their significance for disease progression is less well understood than in other settings.

Perhaps best characterized in animal models is the development of ectopic lymphoid structures associated with salivary gland inoculation of MCMV [63] and in the bronchus-associated tertiary lymphoid tissue associated with influenza virus infection [64]. A recent comparative study of *M. tuberculosis* infection in humans, in non-human primates and in mouse models provides perhaps the best characterized evaluation of the role of ectopic lymphoid tissue in disease progression [65]. Of note, whereas tuberculosis (TB) granulomas in patients and non-human primates with latent disease had associated ectopic lymphoid tissue, this was not the case for granulomas in patient or animals with active TB, suggesting a role for these structures in immune control. The finding of ectopic lymphoid tissue associated with the TB granuloma may be a special case, associated with either the chronicity of infection or the inherent adjuvant properties and/or immunogenicity of this pathogen. In other granulomatous diseases, e.g. experimental visceral leishmaniasis [66], granulomas do not acquire this feature.

2.8 Closure: Stromal Cells and the Resolution of Inflammation

Pathogen clearance ultimately leads to a reversal of most of the associated tissue pathology, through an active process of resolution. Recent evidence suggests that resolution is as complex a process as the generation of immunity and immunopathology, involving a plethora of distinct signals often mediated through shifts in metabolic profile of the tissue. As stromal cells are regarded as the key drivers for chronic inflammation [67], it goes without saying that resolution must bring about changes to the stromal compartment and that stromal cells may indeed drive this process, for example, through consumption of survival signals or the active production of resolution-promoting molecules [68].

The role of pro-resolution inflammatory mediators in the resolution of inflammation during infectious disease, including the role of lipoxins in models of *Toxoplasma gondii*, *Trypanosoma cruzi* and *Plasmodium berghei* infections, has recently been reviewed [69], whereas the role of resolvins has been most clearly illustrated in the control of herpes simplex virus infection [70]. Nevertheless, in the context of infectious disease, this area of stromal cell biology still offers huge potential not only for uncovering new regulatory pathways per se but for the identification of novel approaches to hasten, when appropriate, the resolution of infection-associated inflammation in the clinic.

2.9 Host-Directed Therapy: Stromal-Targeted Therapeutics for Infectious Disease

The development of immunotherapies targeting stromal cells is also described in detail elsewhere in this volume. Studies in infectious disease have provided three main examples to date where stromal cell-targeted immunotherapy might have an impact on the disease outcome. In the case of HIV, there is a good correlation between the extent of FRC disruption, collagen deposition and the ability of HAART to restore CD4[+] Tcell count in patients [54, 71]. In experimental MCMV infection, targeting the LTβR with an agonistic mAb was able to restore otherwise defective CCL21 production and to improve homing of T cells into the T zone of MCMV-infected mice [58]. Finally, in an experimental model of visceral leishmaniasis, therapeutic administration of the broad-spectrum tyrosine kinase inhibitor sunitinib had no direct therapeutic benefit but was able to restore FRC and FDC networks. When used in a sequential therapy regimen with conventional antimonial-based chemotherapy, a marked dose-sparing effect was observed that correlated with enhanced T cell effector function [62]. Given the narrow therapeutic window for many anti-parasite drugs and the common occurrence of lymphadenopathy and/or splenomegaly, these data suggest that immunotherapies targeted at restoring lymphoid tissue architecture or minimizing collateral damage due to fibrosis may have a unique place in the future development of anti-infective therapies.

2.10 Concluding Remarks

Although there has been an explosion in the study of stromal cells in recent years, an appreciation of their role in infectious disease pathogenesis is still in its infancy. Tools are now becoming available to fate map stromal cells under conditions of ongoing infection, conditionally deplete or modify their function and explore their characteristics at a global level, suggesting a rich harvest awaits those who choose to enter this field, bringing with them the diversity of pathogens that have contributed so much in the past in terms of understanding the biology of haematopoietic cells. Ultimately, the ability to study stromal cell populations in human infectious disease will become more tractable, and perhaps in the not too distant future, manipulating stromal cell function to combat infectious disease may become a clinical reality.

Acknowledgements The author thanks his numerous colleagues who have contributed to the study of stromal cell biology in his laboratory and the Medical Research Council and the Wellcome Trust for providing long-term research support. The author also apologizes to the many investigators whose work has been omitted for the sake of brevity but who have set the scene for future studies of the role of stromal cells in infection.

References

1. Donjacour AA, Cunha GR. Stromal regulation of epithelial function. Cancer Treat Res. 1991;53:335–64.
2. Zhang X, Martinez D, Koledova Z, Qiao G, Streuli CH, Lu P. FGF ligands of the postnatal mammary stroma regulate distinct aspects of epithelial morphogenesis. Development. 2014;141(17):3352–62.
3. Espinosa-Cantellano M, Martinez-Palomo A. Pathogenesis of intestinal amebiasis: from molecules to disease. Clin Microbiol Rev. 2000;13(2):318–31.
4. Artis D, Grencis RK. The intestinal epithelium: sensors to effectors in nematode infection. Mucosal Immunol. 2008;1(4):252–64.
5. Hiemstra IH, Klaver EJ, Vrijland K, Kringel H, Andreasen A, Bouma G, Kraal G, van Die I, den Haan JM. Excreted/secreted Trichuris suis products reduce barrier function and suppress inflammatory cytokine production of intestinal epithelial cells. Mol Immunol. 2014;60(1):1–7.
6. Kaye P, Scott P. Leishmaniasis: complexity at the host-pathogen interface. Nat Rev Microbiol. 2011;9(8):604–15.
7. Barrias ES, de Carvalho TM, De Souza W. Trypanosoma cruzi: entry into mammalian host cells and parasitophorous vacuole formation. Front Immunol. 2013;4:186.
8. Bogdan C, Donhauser N, Doring R, Rollinghoff M, Diefenbach A, Rittig MG. Fibroblasts as host cells in latent leishmaniosis. J Exp Med. 2000;191(12):2121–30.
9. Svensson M, Maroof A, Ato M, Kaye PM. Stromal cells direct local differentiation of regulatory dendritic cells. Immunity. 2004;21(6):805–16.
10. Hsu KM, Pratt JR, Akers WJ, Achilefu SI, Yokoyama WM. Murine cytomegalovirus displays selective infection of cells within hours after systemic administration. J Gen Virol. 2009;90(Pt 1):33–43.
11. Baldwin J, Park PJ, Zanotti B, Maus E, Volin MV, Shukla D, Tiwari V. Susceptibility of human iris stromal cells to herpes simplex virus 1 entry. J Virol. 2013;87(7):4091–6.
12. Valyi-Nagy T, Sheth V, Clement C, Tiwari V, Scanlan P, Kavouras JH, Leach L, Guzman-Hartman G, Dermody TS, Shukla D. Herpes simplex virus entry receptor nectin-1 is widely expressed in the murine eye. Curr Eye Res. 2004;29(4–5):303–9.
13. Keele BF, Tazi L, Gartner S, Liu Y, Burgon TB, Estes JD, Thacker TC, Crandall KA, McArthur JC, Burton GF. Characterization of the follicular dendritic cell reservoir of human immunodeficiency virus type 1. J Virol. 2008;82(11):5548–61.
14. Smith BA, Gartner S, Liu Y, Perelson AS, Stilianakis NI, Keele BF, Kerkering TM, Ferreira-Gonzalez A, Szakal AK, Tew JG, Burton GF. Persistence of infectious HIV on follicular dendritic cells. J Immunol. 2001;166(1):690–6.
15. Spiegel H, Herbst H, Niedobitek G, Foss HD, Stein H. Follicular dendritic cells are a major reservoir for human immunodeficiency virus type 1 in lymphoid tissues facilitating infection of CD4+ T-helper cells. Am J Pathol. 1992;140(1):15–22.

16. Thacker TC, Zhou X, Estes JD, Jiang Y, Keele BF, Elton TS, Burton GF. Follicular dendritic cells and human immunodeficiency virus type 1 transcription in CD4+ T cells. J Virol. 2009;83(1):150–8.
17. Baud J, Varon C, Chabas S, Chambonnier L, Darfeuille F, Staedel C. Helicobacter pylori initiates a mesenchymal transition through ZEB1 in gastric epithelial cells. PLoS One. 2013;8(4):e60315.
18. Reilkoff RA, Bucala R, Herzog EL. Fibrocytes: emerging effector cells in chronic inflammation. Nat Rev Immunol. 2011;11(6):427–35.
19. Macedo-Silva RM, Santos Cde L, Diniz VA, Carvalho JJ, Guerra C, Corte-Real S. Peripheral blood fibrocytes: new information to explain the dynamics of *Leishmania* infection. Mem Inst Oswaldo Cruz. 2014;109(1):61–9.
20. Allahverdiyev AM, Bagirova M, Elcicek S, Koc RC, Baydar SY, Findikli N, Oztel ON. Adipose tissue-derived mesenchymal stem cells as a new host cell in latent leishmaniasis. Am J Trop Med Hyg. 2011;85(3):535–9.
21. Schilling JD, Martin SM, Hung CS, Lorenz RG, Hultgren SJ. Toll-like receptor 4 on stromal and hematopoietic cells mediates innate resistance to uropathogenic *Escherichia coli*. Proc Natl Acad Sci U S A. 2003;100(7):4203–8.
22. Spanier J, Lienenklaus S, Paijo J, Kessler A, Borst K, Heindorf S, Baker DP, Kroger A, Weiss S, Detje CN, Staeheli P, Kalinke U. Concomitant TLR/RLH signaling of radioresistant and radiosensitive cells is essential for protection against vesicular stomatitis virus infection. J Immunol. 2014;193(6):3045–54.
23. Tomalka J, Ganesan S, Azodi E, Patel K, Majmudar P, Hall BA, Fitzgerald KA, Hise AG. A novel role for the NLRC4 inflammasome in mucosal defenses against the fungal pathogen *Candida albicans*. PLoS Pathog. 2011;7(12):e1002379.
24. Schneider K, Loewendorf A, De Trez C, Fulton J, Rhode A, Shumway H, Ha S, Patterson G, Pfeffer K, Nedospasov SA, Ware CF, Benedict CA. Lymphotoxin-mediated crosstalk between B cells and splenic stroma promotes the initial type I interferon response to cytomegalovirus. Cell Host Microbe. 2008;3(2):67–76.
25. Verma S, Wang Q, Chodaczek G, Benedict CA. Lymphoid-tissue stromal cells coordinate innate defense to cytomegalovirus. J Virol. 2013;87(11):6201–10.
26. Andersson U, Tracey KJ. HMGB1 is a therapeutic target for sterile inflammation and infection. Annu Rev Immunol. 2011;29:139–62.
27. Chan JK, Roth J, Oppenheim JJ, Tracey KJ, Vogl T, Feldmann M, Horwood N, Nanchahal J. Alarmins: awaiting a clinical response. J Clin Invest. 2012;122(8):2711–9.
28. Gomez Perdiguero E, Klapproth K, Schulz C, Busch K, Azzoni E, Crozet L, Garner H, Trouillet C, de Bruijn MF, Geissmann F, Rodewald HR. Tissue-resident macrophages originate from yolk-sac-derived erythro-myeloid progenitors. Nature. 2015;518(7540):547–51.
29. Schulz C, Gomez Perdiguero E, Chorro L, Szabo-Rogers H, Cagnard N, Kierdorf K, Prinz M, Wu B, Jacobsen SE, Pollard JW, Frampton J, Liu KJ, Geissmann F. A lineage of myeloid cells independent of Myb and hematopoietic stem cells. Science. 2012;336(6077):86–90.
30. Yona S, Kim KW, Wolf Y, Mildner A, Varol D, Breker M, Strauss-Ayali D, Viukov S, Guilliams M, Misharin A, Hume DA, Perlman H, Malissen B, Zelzer E, Jung S. Fate mapping reveals origins and dynamics of monocytes and tissue macrophages under homeostasis. Immunity. 2013;38(1):79–91.
31. Onder L, Narang P, Scandella E, Chai Q, Iolyeva M, Hoorweg K, Halin C, Richie E, Kaye P, Westermann J, Cupedo T, Coles M, Ludewig B. IL-7-producing stromal cells are critical for lymph node remodeling. Blood. 2012;120(24):4675–83.
32. Cohen JN, Guidi CJ, Tewalt EF, Qiao H, Rouhani SJ, Ruddell A, Farr AG, Tung KS, Engelhard VH. Lymph node-resident lymphatic endothelial cells mediate peripheral tolerance via Aire-independent direct antigen presentation. J Exp Med. 2010;207(4):681–8.
33. Fletcher AL, Lukacs-Kornek V, Reynoso ED, Pinner SE, Bellemare-Pelletier A, Curry MS, Collier AR, Boyd RL, Turley SJ. Lymph node fibroblastic reticular cells directly present peripheral tissue antigen under steady-state and inflammatory conditions. J Exp Med. 2010;207(4):689–97.

34. Baptista AP, Roozendaal R, Reijmers RM, Koning JJ, Unger WW, Greuter M, Keuning ED, Molenaar R, Goverse G, Sneeboer MM, den Haan JM, Boes M, Mebius RE. Lymph node stromal cells constrain immunity via MHC class II self-antigen presentation. eLife. 2014; 3:e04433

35. Dubrot J, Duraes FV, Potin L, Capotosti F, Brighouse D, Suter T, LeibundGut-Landmann S, Garbi N, Reith W, Swartz MA, Hugues S. Lymph node stromal cells acquire peptide-MHCII complexes from dendritic cells and induce antigen-specific CD4(+) T cell tolerance. J Exp Med. 2014;211(6):1153–66.

36. Peduto L, Dulauroy S, Lochner M, Spath GF, Morales MA, Cumano A, Eberl G. Inflammation recapitulates the ontogeny of lymphoid stromal cells. J Immunol. 2009;182(9):5789–99.

37. Despars G, Tan J, Periasamy P, O'Neill HC. The role of stroma in hematopoiesis and dendritic cell development. Curr Stem Cell Res Ther. 2007;2(1):23–9.

38. O'Neill HC, Griffiths KL, Periasamy P, Hinton RA, Petvises S, Hey YY, Tan JK. Spleen stroma maintains progenitors and supports long-term hematopoiesis. Curr Stem Cell Res Ther. 2014;9(4):354–63.

39. Tan JK, Periasamy P, O'Neill HC. Delineation of precursors in murine spleen that develop in contact with splenic endothelium to give novel dendritic-like cells. Blood. 2010;115(18):3678–85.

40. Nguyen Hoang AT, Liu H, Juarez J, Aziz N, Kaye PM, Svensson M. Stromal cell-derived CXCL12 and CCL8 cooperate to support increased development of regulatory dendritic cells following *Leishmania* infection. J Immunol. 2010;185(4):2360–71.

41. Li Q, Guo Z, Xu X, Xia S, Cao X. Pulmonary stromal cells induce the generation of regulatory DC attenuating T-cell-mediated lung inflammation. Eur J Immunol. 2008;38(10):2751–61.

42. Tang H, Guo Z, Zhang M, Wang J, Chen G, Cao X. Endothelial stroma programs hematopoietic stem cells to differentiate into regulatory dendritic cells through IL-10. Blood. 2006;108(4):1189–97.

43. Xia S, Guo Z, Xu X, Yi H, Wang Q, Cao X. Hepatic microenvironment programs hematopoietic progenitor differentiation into regulatory dendritic cells, maintaining liver tolerance. Blood. 2008;112(8):3175–85.

44. Xu X, Yi H, Guo Z, Qian C, Xia S, Yao Y, Cao X. Splenic stroma-educated regulatory dendritic cells induce apoptosis of activated CD4 T cells via Fas ligand-enhanced IFN-gamma and nitric oxide. J Immunol. 2012;188(3):1168–77.

45. Zhang M, Tang H, Guo Z, An H, Zhu X, Song W, Guo J, Huang X, Chen T, Wang J, Cao X. Splenic stroma drives mature dendritic cells to differentiate into regulatory dendritic cells. Nat Immunol. 2004;5(11):1124–33.

46. Li L, Liu S, Zhang T, Pan W, Yang X, Cao X. Splenic stromal microenvironment negatively regulates virus-activated plasmacytoid dendritic cells through TGF-beta. J Immunol. 2008;180(5):2951–6.

47. Hussaarts L, van der Vlugt LE, Yazdanbakhsh M, Smits HH. Regulatory B-cell induction by helminths: implications for allergic disease. J Allergy Clin Immunol. 2011;128(4):733–9.

48. Turner JD, Narang P, Coles MC, Mountford AP. Blood flukes exploit Peyer's Patch lymphoid tissue to facilitate transmission from the mammalian host. PLoS Pathog. 2012;8(12):e1003063.

49. Wood GS. The immunohistology of lymph nodes in HIV infection: a review. Prog AIDS Pathol. 1990;2:25–32.

50. Wood GS, Garcia CF, Dorfman RF, Warnke RA. The immunohistology of follicle lysis in lymph node biopsies from homosexual men. Blood. 1985;66(5):1092–7.

51. Rosenberg YJ, Lewis MG, Greenhouse JJ, Cafaro A, Leon EC, Brown CR, Bieg KE, Kosco-Vilbois MH. Enhanced follicular dendritic cell function in lymph nodes of simian immunodeficiency virus-infected macaques: consequences for pathogenesis. Eur J Immunol. 1997;27(12):3214–22.

52. Smelt SC, Engwerda CR, McCrossen M, Kaye PM. Destruction of follicular dendritic cells during chronic visceral leishmaniasis. J Immunol. 1997;158(8):3813–21.

53. Zeng M, Paiardini M, Engram JC, Beilman GJ, Chipman JG, Schacker TW, Silvestri G, Haase AT. Critical role of CD4 T cells in maintaining lymphoid tissue structure for immune cell homeostasis and reconstitution. Blood. 2012;120(9):1856–67.

54. Zeng M, Southern PJ, Reilly CS, Beilman GJ, Chipman JG, Schacker TW, Haase AT. Lymphoid tissue damage in HIV-1 infection depletes naive T cells and limits T cell reconstitution after antiretroviral therapy. PLoS Pathog. 2012;8(1):e1002437.
55. Ato M, Stager S, Engwerda CR, Kaye PM. Defective CCR7 expression on dendritic cells contributes to the development of visceral leishmaniasis. Nat Immunol. 2002;3(12):1185–91.
56. Graw F, Regoes RR. Influence of the fibroblastic reticular network on cell-cell interactions in lymphoid organs. PLoS Comput Biol. 2012;8(3):e1002436.
57. Scandella E, Bolinger B, Lattmann E, Miller S, Favre S, Littman DR, Finke D, Luther SA, Junt T, Ludewig B. Restoration of lymphoid organ integrity through the interaction of lymphoid tissue-inducer cells with stroma of the T cell zone. Nat Immunol. 2008;9(6):667–75.
58. Benedict CA, De Trez C, Schneider K, Ha S, Patterson G, Ware CF. Specific remodeling of splenic architecture by cytomegalovirus. PLoS Pathog. 2006;2(3):e16.
59. Engwerda CR, Ato M, Cotterell SE, Mynott TL, Tschannerl A, Gorak-Stolinska PM, Kaye PM. A role for tumor necrosis factor-alpha in remodeling the splenic marginal zone during Leishmania donovani infection. Am J Pathol. 2002;161(2):429–37.
60. Beattie L, Engwerda CR, Wykes M, Good MF. CD8+ T lymphocyte-mediated loss of marginal metallophilic macrophages following infection with Plasmodium chabaudi chabaudi AS. J Immunol. 2006;177(4):2518–26.
61. Yurdakul P, Dalton J, Beattie L, Brown N, Erguven S, Maroof A, Kaye PM. Compartment-specific remodeling of splenic micro-architecture during experimental visceral leishmaniasis. Am J Pathol. 2011;179(1):23–9.
62. Dalton JE, Maroof A, Owens BM, Narang P, Johnson K, Brown N, Rosenquist L, Beattie L, Coles M, Kaye PM. Inhibition of receptor tyrosine kinases restores immunocompetence and improves immune-dependent chemotherapy against experimental leishmaniasis in mice. J Clin Invest. 2010;120(4):1204–16.
63. Grewal JS, Pilgrim MJ, Grewal S, Kasman L, Werner P, Bruorton ME, London SD, London L. Salivary glands act as mucosal inductive sites via the formation of ectopic germinal centers after site-restricted MCMV infection. FASEB J. 2011;25(5):1680–96.
64. GeurtsvanKessel CH, Willart MA, Bergen IM, van Rijt LS, Muskens F, Elewaut D, Osterhaus AD, Hendriks R, Rimmelzwaan GF, Lambrecht BN. Dendritic cells are crucial for maintenance of tertiary lymphoid structures in the lung of influenza virus-infected mice. J Exp Med. 2009;206(11):2339–49.
65. Slight SR, Rangel-Moreno J, Gopal R, Lin Y, Fallert Junecko BA, Mehra S, Selman M, Becerril-Villanueva E, Baquera-Heredia J, Pavon L, Kaushal D, Reinhart TA, Randall TD, Khader SA. CXCR5(+) T helper cells mediate protective immunity against tuberculosis. J Clin Invest. 2013;123(2):712–26.
66. Moore JW, Beattie L, Dalton JE, Owens BM, Maroof A, Coles MC, Kaye PM. B cell: T cell interactions occur within hepatic granulomas during experimental visceral leishmaniasis. PLoS One. 2012;7(3):e34143.
67. Barone F, Nayar S, Buckley CD. The role of non-hematopoietic stromal cells in the persistence of inflammation. Front Immunol. 2012;3:416.
68. Serhan CN, Brain SD, Buckley CD, Gilroy DW, Haslett C, O'Neill LA, Perretti M, Rossi AG, Wallace JL. Resolution of inflammation: state of the art, definitions and terms. FASEB J. 2007;21(2):325–32.
69. Russell CD, Schwarze J. The role of pro-resolution lipid mediators in infectious disease. Immunology. 2014;141(2):166–73.
70. Rajasagi NK, Reddy PB, Suryawanshi A, Mulik S, Gjorstrup P, Rouse BT. Controlling herpes simplex virus-induced ocular inflammatory lesions with the lipid-derived mediator resolvin E1. J Immunol. 2011;186(3):1735–46.
71. Zeng M, Haase AT, Schacker TW. Lymphoid tissue structure and HIV-1 infection: life or death for T cells. Trends Immunol. 2012;33(6):306–14.

Chapter 3
Fibroblasts and Osteoblasts in Inflammation and Bone Damage

Jason D. Turner, Amy J. Naylor, Christopher Buckley, Andrew Filer, and Paul-Peter Tak

Abstract This review discusses the important role fibroblasts play in the process of inflammation and the evidence that these cells may drive the persistence of inflammation. Fibroblasts are key components of the stroma normally involved in maintenance of extracellular matrix and tissue function; however, the term 'fibroblast' is used to describe a heterogeneous population of cells that vary in phenotype both between and within anatomical sites. Fibroblasts possess Toll-like receptors allowing them to respond to pathogen and damage-related signals by producing proinflammatory mediators such as IL-6, PGE_2, and GM-CSF and can produce a range of chemokines such as CXCL12, CXCL13, and CXCL8 which attract B and T lymphocytes, monocytes, and neutrophils to sites of inflammation. Interactions between leukocytes and fibroblasts can facilitate increased survival of the leukocytes and modulate phenotypes leading to differential gene expression in the presence of mediators involved in inflammation. Fibroblasts also contribute to collateral tissue damage during inflammation through the production of members of the metalloproteinase family and cathepsins and also through induction of osteoclastogenesis leading to increased bone resorption rates. In persistent diseases, fibroblasts obtain an imprinted, aggressive phenotype leading to the production of higher basal levels of proinflammatory cytokines and the ability to damage tissue in the absence of continual stimuli. This aggressive phenotype offers an attractive new target for therapeutics that could help alleviate the burden of persistent inflammation.

J. D. Turner (✉) · A. J. Naylor · A. Filer
Rheumatology Research Group, Institute for Inflammation and Ageing, College of Medical and Dental Sciences, University of Birmingham, Queen Elizabeth Hospital, Birmingham, UK
e-mail: j.d.turner@bham.ac.uk

C. Buckley
Rheumatology Research Group, Institute for Inflammation and Ageing, College of Medical and Dental Sciences, University of Birmingham, Queen Elizabeth Hospital, Birmingham, UK

The Kennedy Institute of Rheumatology, University of Oxford, Oxford, UK

P.-P. Tak
Division of Clinical Immunology & Rheumatology, Academic Medical Center/University of Amsterdam, Amsterdam, The Netherlands

© Springer International Publishing AG, part of Springer Nature 2018
B. M.J Owens, M. A. Lakins (eds.), *Stromal Immunology*, Advances in Experimental Medicine and Biology 1060, https://doi.org/10.1007/978-3-319-78127 3_3

Keywords Fibroblast · Synovial fibroblast · Osteoblast · Rheumatoid arthritis · Stromal immunology · Epigenetics

3.1   Introduction

Historically inflammation has been considered to be a process driven by leukocytes, with tissues and the stromal cells within viewed as bystanders in inflammatory progression and resolution. Such cells may include tissue fibroblasts, endothelial cells, and supportive cells such as pericytes. It is now apparent that the stroma is critically involved in all stages of inflammation and may play a role in the switch from resolving to persistent inflammation. Researchers investigating diseases such as cancer, fibrosis, and rheumatoid arthritis (RA) are focusing efforts on elucidating the role the stroma plays in inflammation. In particular, a key cellular component of the stroma, the fibroblast, has been incriminated in multiple diseases and shows promise as a target of future therapeutics. Another tissue-specific stromal component of mesenchymal origin, the osteoblast, plays an important role in inflammation in rheumatoid arthritis (RA), a systemic persistent inflammatory disorder that will be used as a model of the fibroblast inflammatory phenotype in this review.

3.2   What is a Fibroblast?

The term 'fibroblast' is used to describe organ-specific resident cells of body tissues and organs whose primary function is to maintain the extracellular matrix (ECM) of those tissues in health and during wound healing. Although sharing the common title of fibroblast, the function and phenotype of these cells are specialised towards the site in which they reside.

3.2.1   Fibroblast Origins

It is accepted that fibroblasts develop from the primary mesenchyme during embryogenesis; however, the contribution of blood-borne fibrocytes and mesenchymal stem cells (MSCs) of bone marrow origin to the fibroblast pool during wound healing and inflammation is an area of contention [1].

Fibrocytes have been identified in the pool of peripheral blood mononuclear cells (PBMCs) in humans and mice [2]. They are unusual as they initially express the haematopoietic marker CD45 which fibroblasts lack but also express markers associated with fibroblasts and wound healing such as collagen I and αSMA and have the ability to contract ECM. Treatment of PBMCs with transforming growth factor-β1 (TGF-β1) can drive fibrocyte differentiation from a CD14+ precursor highlighting the link these cells have to tissue maintenance and repair [2, 3]. In addition fibrocytes express a range of chemokine receptors such as CXCR4, CCR3, CCR5, and CCR7 that facilitate migration to wounds as demonstrated using a murine skin puncture model. However, it remains unclear whether fibrocytes actually give rise to tissue fibroblasts or instead serve a specialised role during wound healing.

MSCs have been isolated from the majority of connective tissues and are identified using a broad range of markers such as the presence of CD73, CD90, and CD105 and the absence of markers of other cell populations such as CD45, CD11b, CD19, and HLA-DR (for a comprehensive review, see Uccelli et al. [4]). Treating MSCs with connective tissue growth factor (CTGF) over 2–4 weeks results in a downregulation of markers specific for MSCs and a concurrent upregulation of markers associated with fibroblasts such as vimentin, fibroblast-specific protein-1 (FSP-1), and collagen I and additionally a reduction in the capability of the cells to differentiate into other MSC-derived lineages such as osteoblasts or adipocytes [5].

3.2.2 Fibroblasts from Different Sites in the Body Are Heterogeneous

Fibroblasts are not a homogenous population and show variation in gene expression and behaviour depending on the site from which the cells are isolated. Microarray analysis of genes associated with inflammation highlighted that not only do unstimulated fibroblasts have differing gene expression patterns but also that the response of fibroblasts to various stimuli varies with the site from which the cells were taken [6]. Filer et al. [7] demonstrated that differentially expressed genes in dermal, synovial, and bone marrow fibroblasts follow a hierarchy with the largest number of differentially expressed genes being between anatomical locations, followed by response to serum and finally disease (RA vs. osteoarthritis (OA)). Although fibroblasts from different sites share generic aspects of the serum response programme and the effect of disease, site-specific differences unique to each tissue of origin were also observed.

Variation in fibroblast phenotype is not limited to fibroblasts from different tissues but also occurs between fibroblasts taken from different sites within the same tissue. Comparing transcriptomes of predominantly dermal fibroblasts from 43 anatomical locations, Rinn et al. [8] demonstrated clustering of fibroblasts from the same geographic region of the body and that the variance within sites is less than that between donors, a finding which had previously been reported by Chang et al. [9]. Some of the differentially expressed genes are members of the HOX gene family which are involved in embryological patterning and development. This is also evidenced by maintenance of the site-specific gene patterns even after long-term culture.

Work in this area led to the hypothesis of a stromal address code that functions in concert with endothelial tissues [1]. This address code, mediated via chemokine and adhesion molecule expression, regulates the influx and efflux of appropriate leukocytes from the endothelium into tissues or from tissues into lymphatics. This theory also provides an interesting perspective on persistent inflammatory diseases, viewing the persistence of inflammation as a result of an inappropriate shift in stromal postcode towards a lymphatic pattern, rather than solely an effect of the inflammatory milieu.

3.2.3 *Stromal Subpopulations Exist Within Tissues*

In addition to the variation of fibroblast phenotype between sites, it is apparent that fibroblasts vary within sites and various subsets/phenotypes can be identified. Using the synovium as an example which has been extensively studied in RA, at least two phenotypes of fibroblast can be identified. The synovium can be segregated into the lining layer, which is adjacent to the joint space, and the less organised sublining layer. Lining layer fibroblasts can be identified by the expression of cadherin-11, a cell surface marker that allows homotypic adhesion of the lining layer fibroblasts to one another facilitating the formation of a functional lining layer in the absence of a basal lamina [10–12]. Other markers have also been associated with lining layer fibroblasts such as CD55, fibroblast activation protein (FAP), and podoplanin (GP38) [13–15]. Sublining fibroblasts can be identified with alternative markers such as CD90 (THY1) or CD248 (endosialin) and appear to have different roles to lining layer fibroblasts [13, 16–19].

3.3 Regulation of Inflammation

3.3.1 *Fibroblasts Can Both Respond to and Promote Inflammation*

A sign that fibroblasts are not merely bystanders during the course of inflammation but can actually respond to proinflammatory signals can be seen in the capability of synovial fibroblasts to respond to IL-1β and TNFα with an increased proliferative rate [20]. This proliferation is not unrestrained as IL-1β and TNFα also induce the expression of prostaglandin E2 from the cells and this mediator inhibits proliferation demonstrating autocrine regulation of fibroblast proliferation in response to two prototypical proinflammatory mediators.

However, proliferation alone is not truly demonstrative of involvement in inflammation. Synovial fibroblasts express multiple Toll-like receptors (TLRs) and so possess the ability to respond to pathogens or damage casting them in an immune sentinel role with similarities to macrophages [21]. TLR3 and TLR4 are the most abundantly expressed TLRs in synovial fibroblasts, but fibroblasts can also respond to PAMPs such as flagellin and bacterial lipoprotein through TLR5 and TLR1 or TLR6, respectively. Poly(I:C), LPS, bacterial lipoprotein, and flagellin all stimulate production of IL-6 in synovial and, to a lesser extent, skin fibroblasts. Synovial fibroblasts isolated from the joints of patients with RA (RASF) express mRNA for IL-6, IL-11, and OSM and in response to IL-1α or TNFα increase the production of these cytokines [22]. RASF can also respond to IL-17 stimulation with increased secretion of IL-6 and the neutrophil chemoattractant CXCL8 [23], whilst TNFα and T-cell-derived IL-17 induce synergistic production of GM-CSF and neutrophil survival [23, 24]. Additionally, lung fibroblasts have been shown to produce GM-CSF in response to IL-1α or IL-1β allowing them to influence the differentiation and survival of cells such as monocytes [25].

Fibroblasts can regulate the response of a large number of TNFα-responsive genes in macrophages. In vitro coculture of M-CSF differentiated macrophages with synovial or lung fibroblasts in the presence of TNFα results in differential regulation of genes in macrophages that are normally up- or downregulated by TNFα [26]. The upregulation of around 22% of genes by TNFα was attenuated by ≥50% during coculture compared to macrophages cultured alone, and interestingly many attenuated genes were related to interferon- or *myc*-regulated gene signatures, indicating a coordinated response to coculture. Conversely the expression of approximately 34% of genes normally downregulated in macrophages by TNFα was upregulated by twofold or more in the presence of fibroblasts, with the genes affected related to growth factors such as TGF-β, M-CSF, and GM-CSF. Prior to this study, it was known that fibroblast and monocyte coculture indirectly increases the release of IL-6 in an IL-1β-dependent manner and that GM-CSF, LIF, and CXCL8 are also increased simply through the coculture of these two cell types [27].

Taking these findings together highlights not only the involvement of fibroblasts in responding to inflammatory cues but also the ability of this family of cells to produce mediators involved in the process of inflammation and even manipulate the response of cells of the immune system to pro-inflammatory mediators.

3.3.2 Fibroblasts Regulate Leukocyte Infiltration and Survival

The ability of fibroblasts to regulate the recruitment or retention of leukocytes within tissues is well documented. RASF constitutively express CXCL12, CCL2, and CXCL8 which through interactions with CXCR4, CCR2, CCR4, CXCR1, and CXCR2 facilitate the infiltration of B and T lymphocytes, monocytes, and neutrophils [28–30]. RASF that have been stimulated with the TLR2 ligand peptidoglycan increase the secretion of CXCL8, CCL5, and CCL8 which are chemoattractants of neutrophils, monocytes, and CD4[+] T cells [31]. The concentration of CXCL12 in RA synovial fluid is higher than that of patients with OA. Coculture of RASF with CD4[+] T cells increases CXCL12 production via CD40-CD40L interactions and IL-17 release [32]. Furthermore, RASF secrete higher levels of CCL2 and CXCL8 than synovial fibroblasts from osteoarthritis patients (OASF) or dermal fibroblasts leading to increased levels of monocyte migration [33].

Inappropriate retention of leukocyte subsets within a tissue leads to persistence of inflammation. Retention, as modelled by assays of pseudoemperipolesis, in which a monolayer of stromal cells facilitates migration of cells underneath the monolayer, is often used as an in vitro assessment of the ability of fibroblasts to hold cells within tissues. RASF and OASF induce pseudoemperipolesis in peripheral blood B cells and activated T cells, whereas dermal fibroblasts can only induce limited B-cell pseudoemperipolesis and do not have this effect on mature T cells [28, 29]. In B cells this process is mediated via CXCL12-CXCR4 and VCAM-1-VLA-4 interactions, whereas in T cells only the CXCL12-CXCR4 axis is required. Synovial fibroblasts can also support the pseudoemperipolesis of natural killer (NK) cells, and coculture of the two cell types together elicits an increase in IL-15, GM-CSF, IL-6, CXCL8, and CCL2 [34].

The survival of leukocytes is also increased by coculture with synovial fibroblasts. Dermal fibroblasts and OASF increase the viability of B cells after isolation but not to the same extent as RASF which increase viability from 3.6% in monoculture to 53.2% in coculture after 6 days [29, 35]. RASF constitutively express membrane-bound IL-15 and B-cell activating factor (BAFF), and TLR3 ligation can increase BAFF and a proliferation-inducing ligand (APRIL) expression, whereas IL-15 expression is increased in synovial fibroblast-NK cell cocultures [34–36]. BAFF increases the expression of IL-15R which through engagement with IL-15 acts as a survival signal for B cells and NK cells in combination with cell contact. In addition to promoting the survival and retention of B cells, BAFF and APRIL also induce class switching in B cells, demonstrated by the induction of activation-induced cytidine deaminase (AID) and an increase in IgA and IgG expression, indicating fibroblasts can affect the differentiation of B cells. Activated CD4 T-cell survival is also increased through coculture with synovial fibroblasts in an IFNβ-dependent method [37].

3.3.3 Inflammation Can Drive Aberrant Expression of ECM Remodelling Enzymes

RASF are known to produce ECM remodelling factors such as MMP3, MMP9, and MMP13 and cathepsins B, D, and L [21, 34, 38, 39]. The mediators produced facilitate the invasion of RASF into the cartilage compared to limited invasion by OASF that have not been involved in the persistent inflammation seen in RA [38, 40]. RASF are capable of invading the cartilage in the absence of stimuli from leukocytes confirming that in persistent disease fibroblasts can obtain an 'imprinted' aggressive phenotype and, in an in vivo model of cartilage invasion (severe combined immunodeficiency mouse cartilage xenograft model), fibroblasts can migrate from the cartilage in one site to another via the vasculature offering hints towards the temporal involvement of more joints in RA [41].

The formation of fibroblast-rich pannus tissue that invades and damages the cartilage and bone is a signature of RA. The invasive fibroblasts within the pannus tissue most likely derive from the subset of lining layer fibroblasts within the synovium given that cadherin-11 is expressed in invading pannus and up- or down-regulation of this marker has a corresponding effect upon invasion of the cartilage [10, 11]. There is also evidence suggesting that the interaction of fibroblasts with cells of the monocyte/macrophage lineage can increase the invasive capability of these cells in vitro via increased MMP production [42–44].

3.3.4 Histone Methylation and Acetylation Are Perturbed in RASF

RASF are characterised by a persistent pro-inflammatory phenotype. Epigenetic mechanisms regulating accessibility of DNA transcription complexes to gene promoters are thought to play a key role in maintenance of this phenotype and include modifications of the histone proteins around which DNA is wound, such as acetylation, methylation and phosphorylation, and direct methylation of CpG dinucleotides of DNA (for a review see Portela and Esteller [45]). Changes to histone proteins are regulated by protein complexes containing enzymes that add or remove groups at specific amino acid residues, for instance, acetyl groups are added and removed, respectively, by families of histone acetyltransferases (HATs) and histone deacetylases (HDACs). Methyl groups are added to CpG dinucleotides within DNA by a well-described family of DNA methyltransferases (DNMTs), although mechanisms of dynamic removal of methylation groups remain poorly understood [46].

Results regarding the activity of enzymes involved in histone acetylation are contrasting. Huber et al. [47] found that nuclear HDAC activity is significantly lower in extracts from RA synovial tissue than from osteoarthritis or normal synovial tissue, whereas Kawabata et al. [48] found evidence for the opposite. Both studies agreed that there was no significant difference in the activity of HATs. Of interest, Kawabata et al. [48] proceeded to investigate the levels of HDAC mRNA and found that levels of HDAC1 were significantly higher in RA synovium than controls. Stimulation of RASF with TNFα increases HDAC1 mRNA and HDAC activity, and TNFα and HDAC1 transcript levels are positively correlated. Subsequent studies have also provided contrasting findings with histone H3 acetylation in the IL-6 promoter found at higher levels in RASF than OASF, IL-6 mRNA being expressed at higher basal levels in RASF, and the inhibition of HATs decreasing both histone H3 acetylation and IL-6 production in response to TNFα [49]. On the other hand, Grabiec et al. [50] found that inhibition of HDAC enzymes inhibited IL-6 production in response to TNFα which highlights the complexity associated with histone marks and the possibility of both activation and inhibition of gene expression depending on the location of the modification.

Levels of DNA methylation also vary in RA compared to osteoarthritis. RASF are hypomethylated compared to OASF within the synovial tissue and maintain this difference during in vitro culture [51]. Treating synovial fibroblasts from healthy donors with the demethylating agent 5-azacytidine for a period of 3 months upregulates around 180 genes by more than twofold, many of which have been implicated in RA. When comparing RASF and OASF, a different study found 575 genes associated with hypomethylated sites and 714 genes associated with hypermethylated sites with a total of 3470 differentially expressed genes highlighting the changes in fibroblast phenotype elicited by the RA environment [52].

3.3.5　MicroRNAs Regulate Fibroblast Biology

MicroRNAs (miRNAs) are small non-coding RNA sequences 21–25 nucleotides in length that regulate the expression and stability of multiple coding mRNA species through direct and indirect mechanisms (for a comprehensive review, see He and Hannon [53]). The coordinated expression of miRNAs is coming to prominence as an important aspect of the induction and resolution of inflammation.

RASF have increased basal expression of several miRNAs compared to OASF such as miR-155 and miR-146a. The expression of miR-155 in RA synovial tissue is eightfold higher than in OA [54, 55]. miR-155 acts to downregulate the expression of the matrix-destructive enzymes matrix-metalloproteinase-3 (MMP3) and MMP1 in response to various TLR ligands and IL-1β, indicating that increased expression of miR-155 may be a pro-resolution regulator released during inflammation. Another miRNA, miR-22, downregulates proliferation and IL-6 production by synovial fibroblasts through downregulation of the mediator Cry61 and has been found to be expressed at lower levels in the RA synovium than OA [56]. Many miRNAs were found to be differentially regulated between RASF and OASF in a study by de la Rica et al. [52]. The regulatory hierarchy between DNA methylation and miRNA expression was found to be complex, varying from gene to gene.

3.4　Bone Remodelling in RA

In addition to the direct involvement of fibroblasts in cartilage degradation, they can also indirectly regulate bone damage through regulation of the osteoclast/osteoblast axis.

3.4.1　Normal Bone Maintenance Is Perturbed in RA

In order to maintain its strength and integrity, bone tissue is continuously turned over throughout adult life at a rate of approximately 10% total bone content per year. Two key cell types required for this process are the osteoblast, which produces bone, and the osteoclast, which resorbs it. These cells signal to each other via osteoblast production of 'receptor activator of NF-κβ ligand' (RANKL) and osteoprotegerin (OPG). RANKL binds to its receptor RANK on the osteoclast to induce osteoclastogenesis; OPG acts as a RANKL decoy receptor and thus inhibits osteoclast formation (reviewed in Bar-Shavit [57]). This cellular crosstalk (often referred to as osteoblast-osteoclast coupling) serves to balance the activity of the two cell types ensuring equilibrium between bone production and bone resorption is maintained (reviewed in [58]). Defects in this relationship can lead to disorders of bone destruction by osteoclasts (e.g. osteoporosis) or excessive bone formation by osteoblasts (e.g. osteopetrosis).

The coupling between osteoclast-mediated bone resorption and osteoblast-mediated bone formation is perturbed during persistent inflammation. In the RA inflammatory environment, this results in net bone loss manifested in three ways: focal bone loss caused by erosions at the joint margins, periarticular osteopenia in bones adjacent to inflamed joints, and a generalised osteoporosis of the entire skeleton [59]. Despite improvements in RA treatment made since the introduction of biologics and increases in reported rates of remission [60], patients with inadequately controlled RA have increased fracture risk and inadequate fracture healing (reviewed in Claes et al. [61]) making the control of bone integrity an important clinical issue [60].

Osteoblasts are stromal cells derived from MSC precursors, whilst osteoclasts are multinucleated cells of the haematopoietic lineage. Much of the coupling imbalance seen in RA can be explained by defects in osteoblast signalling and activity. The differentiation process from MSC through pre-osteoblast to mature osteoblast is dependent initially on the transcription factor RUNX2 and, as the osteoblast matures, on the combination of RUNX2 and osterix (reviewed in Long [62]). The mature osteoblast produces osteocalcin, alkaline phosphatase, and collagen type I in order to lay down 'osteoid' ECM, which is later mineralised through the accumulation of hydroxyapatite to form bone.

Walsh et al. [63] have shown that in mice with inflammatory arthritis, the rate of osteoid (unmineralised bone matrix) formation by osteoblasts is the same at sites affected by arthritis, where active bone resorption is taking place, as it is at unaffected sites. This alone suggests that the amount of bone formation at sites where active resorption is taking place cannot equal the greatly increased degree of bone loss. Even more strikingly, the degree of mineralised bone formation at sites adjacent to inflammation is reduced compared to non-inflamed sites. A paucity of mature osteoblasts (those expressing alkaline phosphatase) was observed despite the presence of reasonable numbers of immature osteoblasts (cells expressing Runx2) [63].

3.4.2 Pro-inflammatory Cytokines and the Inflammatory Microenvironment Suppress Bone Formation

The cause of this defect may be, at least in part, due to the presence of high levels of pro-inflammatory cytokines during inflammation. Gilbert et al. [64, 65] have identified that addition of TNFα to pre-osteoblast cultures arrests osteoblast differentiation and maturation in vitro. Others have similarly demonstrated that markers of osteoblast maturation such as alkaline phosphatase, collagen type I, and osteocalcin are all reduced in the presence of TNFα and treated cells are unable to upregulate matrix mineralisation [66–69]. Osteoblasts in vitro cultured with serum from patients treated with infliximab (an anti-TNF biologic agent) show reduced expression of IL-6, a cytokine that has been linked to arthritis-related bone loss at least in part by binding the IL-6 receptor, an interaction which induces prostaglandin E2 synthesis, in turn reducing the ratio of OPG/RANKL expression by the osteoblast, favouring osteoclastogenesis [70, 71]. In addition to its effect on IL-6, osteoblasts cultured with serum from patients treated with infliximab show reduced expression of IL-1β, known to inhibit bone formation in vitro and to impair osteoblast migration towards chemotactic factors in vivo [71–74].

In rheumatoid arthritis, the local microenvironment is profoundly changed due to the influx of immune cells and proliferation of synovial fibroblasts within affected joints. This produces a localised hypoxia and a reduced pH, both of which are capable of influencing osteoblasts within the joint. Hypoxia inhibits Wnt signalling (discussed in more detail below) in osteoblasts by sequestering β-catenin to inhibit transcriptional activity and by upregulating DKK-1; low pH causes the downregulation of alkaline phosphatase synthesis in osteoblasts which prevents mineralisation [75–77].

RASF also have the capacity to manipulate the balance of osteoblast/osteoclast directly through the expression of RANKL. In the RA synovium, RANKL expression is higher than in OA, and the expression co-localises with the lining layer marker CD55 [78, 79]. Stimulating RASF with TLR2, TLR3, or TLR4 ligands induces RANKL expression indirectly through the induction of IL-1β expression; however, OASF or dermal fibroblasts do not increase RANKL expression in response to these ligands. RASF stimulated with peptidoglycan, lipopolysaccharide, or poly(I:C) are able to induce osteoclastogenesis in monocytes as demonstrated by the expression of the osteoclast marker tartrate-resistant acid phosphatase (TRAP).

Further evidence that total resolution of inflammation (and thus maximal reduction in the expression of pro-inflammatory factors such as IL-1 and TNFα) is required for recovery of bone integrity comes from mouse studies in which resolving models of RA can be utilised. In one example, Matzelle et al. [80] showed that complete resolution of inflammation allowed for osteoblast-mediated bone formation and repair of bone damage in a process mediated by the induction of the anabolic, pro-mineralisation factors Wnt10b and DKK2 and suppression of Wnt antagonists. TNFα may also inhibit normal osteoblast function through other mechanisms, including effects on the Wnt signalling pathway. Diarra et al. [81] have shown that TNFα modulates

Wnt signalling causing enhanced DKK-1 expression in synovial fibroblasts, whilst blockade of DKK-1 induces fusion of sacroiliac joints mimicking ankylosing spondylitis [82]. Sclerostin, a Wnt signalling inhibitor expressed by osteocytes (osteoblasts that have become entombed within the bone matrix), has shown promise as a drug target in RA. Anti-sclerostin antibodies were able to inhibit bone loss (systemic, periarticular, and local) in mouse models of arthritis [83]. Importantly, this antibody was also able to induce bone repair but only if used in combination with anti-TNF antibody (infliximab), again suggesting that bone repair can only occur when systemic inflammation is controlled.

3.5 Stromal Cells Are Promising Therapeutic Targets

With increasing understanding of the role of stromal cells in inflammation, new therapeutic approaches that differ from the approach of directly targeting inflammatory mediators are being piloted. Treatments with histone deacetylase inhibitors have been used in both mice and humans with positive effects seen in both cases. For example, treatment of human RASF in vitro with the histone deacetylase inhibitor Trichostatin A inhibits the cell cycle and sensitises cells to apoptosis induction by the TNF-related apoptosis-inducing ligand (TRAIL) which is found at higher concentrations in RA synovial fluid than OA [84]. Use of the same treatment in murine collagen antibody-induced arthritis reduced the overall clinical arthritis score, significantly reduced histological signs of synovial inflammation and cartilage damage, and reduced the expression of MMP3 and MMP13 in chondrocytes, cells responsible for the maintenance of the cartilage [85]. Another histone deacetylase inhibitor, givinostat, has been used to treat patients with juvenile idiopathic arthritis and resulted in a reduction in swollen and tender joint counts [86]. However, due to the non-specific nature of the drug, many adverse effects were seen such as vomiting, nausea, and fatigue which may limit the usefulness of the relatively non-selective histone deacetylase inhibitors.

Recent work has indicated miRNA-orientated treatments could be used in arthritic disorders. Treatment of OASF with the miRNA-146a-upregulating compound denbinobin indirectly reduced monocyte adhesion to these cells by interfering with an IL-1β-mediated increase of ICAM-1 and VCAM-1 [87]. Denbinobin also increased HAT activity in OASF. Yao et al. [88] developed a pre-miR-146a delivery system using virus-like particles to upregulate miR-146a expression in monocytes. Upregulation of miR-146a in monocytes inhibited osteoclastogenesis induced by RANKL and M-CSF and reduced bone resorption in vitro.

3.6 Conclusions

Fibroblasts are not merely bystanders during inflammation and can produce and respond to inflammatory mediators (Fig. 3.1). Additionally, they possess TLRs allowing them to respond to pathogens and damage and are critically involved in regulating the influx and retention of leukocytes within tissues. Fibroblasts can be driven to damage the tissues within which they reside through the inappropriate expression of digestive enzymes such as MMPs and also indirectly via regulation of the osteoblast/osteoclast axis. In RA the proinflammatory milieu interferes with normal osteoblast function allowing osteoclast-mediated bone resorption to predominate. Given the roles stromal cells play in inflammation and its persistence, these cells are promising targets for new therapies.

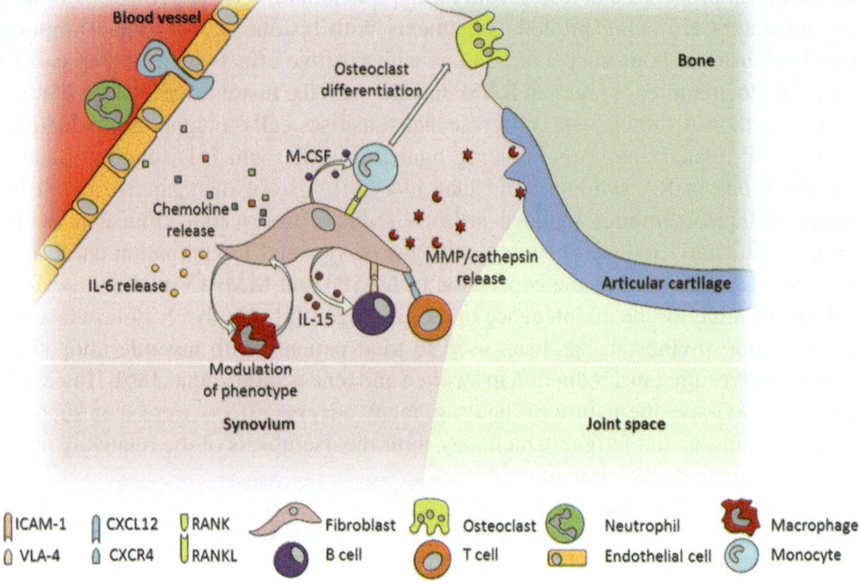

Fig. 3.1 Fibroblasts are heavily involved in inflammation. Chemokines such as CCL2, CXCL8, CXCL12, and CXCL13 attract leukocytes into the tissue where fibroblasts release factors such as IL-6, IL-11, and GM-CSF to propagate inflammation. Interactions of B cells and T cells with fibroblasts via CXCL12-CXCR4 and VCAM-1-VLA-4 interactions retain the cells in the tissue and provide survival signals in concert with IL-15 released from fibroblasts. Binding of fibroblast-expressed RANKL to RANK on monocytes in combination with M-CSF drives osteoclast differentiation. MMP and cathepsin release damages the extracellular matrix and articular cartilage in joints

References

1. Parsonage G, Filer AD, Haworth O, Nash GB, Rainger GE, Salmon M, Buckley CD. A stromal address code defined by fibroblasts. Trends Immunol. 2005;26(3):150–6. https://doi.org/10.1016/j.it.2004.11.014.
2. Abe R, Donnelly SC, Peng T, Bucala R, Metz CN. Peripheral blood fibrocytes: differentiation pathway and migration to wound sites. J Immunol. 2001;166(12):7556–62.
3. Pilling D, Fan T, Huang D, Kaul B, Gomer RH. Identification of markers that distinguish monocyte-derived fibrocytes from monocytes, macrophages, and fibroblasts. PLoS One. 2009;4(10):e7475. https://doi.org/10.1371/journal.pone.0007475.
4. Uccelli A, Moretta L, Pistoia V. Mesenchymal stem cells in health and disease. Nat Rev Immunol. 2008;8(9):726–36. https://doi.org/10.1038/nri2395.
5. Lee CH, Shah B, Moioli EK, Mao JJ. CTGF directs fibroblast differentiation from human mesenchymal stem/stromal cells and defines connective tissue healing in a rodent injury model. J Clin Invest. 2010;120(9):3340–9. https://doi.org/10.1172/jci43230.
6. Parsonage G, Falciani F, Burman A, Filer A, Ross E, Bofill M, Martin S, Salmon M, Buckley CD. Global gene expression profiles in fibroblasts from synovial, skin and lymphoid tissue reveals distinct cytokine and chemokine expression patterns. Thromb Haemost. 2003;90(4):688–97. https://doi.org/10.1267/thro03040688.
7. Filer A, Antczak P, Parsonage GN, Legault HM, O'Toole M, Pearson MJ, Thomas AM, Scheel-Toellner D, Raza K, Buckley CD, Falciani F. Stromal transcriptional profiles reveal hierarchies of anatomical site, serum response and disease and identify disease specific pathways. PLoS One. 2015;10(3):e0120917. https://doi.org/10.1371/journal.pone.0120917.
8. Rinn JL, Bondre C, Gladstone HB, Brown PO, Chang HY. Anatomic demarcation by positional variation in fibroblast gene expression programs. PLoS Genet. 2006;2(7):e119. https://doi.org/10.1371/journal.pgen.0020119.
9. Chang HY, Chi J-T, Dudoit S, Bondre C, Mvd R, Botstein D, Brown PO. Diversity, topographic differentiation, and positional memory in human fibroblasts. Proc Natl Acad Sci U S A. 2002;99(20):12877–82. https://doi.org/10.2307/3073316.
10. Kiener HP, Niederreiter B, Lee DM, Jimenez-Boj E, Smolen JS, Brenner MB. Cadherin 11 promotes invasive behavior of fibroblast-like synoviocytes. Arthritis Rheum. 2009;60(5):1305–10. https://doi.org/10.1002/art.24453.
11. Lee DM, Kiener HP, Agarwal SK, Noss EH, Watts GFM, Chisaka O, Takeichi M, Brenner MB. Cadherin-11 in synovial lining formation and pathology in arthritis. Science. 2007;315(5814):1006–10. https://doi.org/10.1126/science.1137306.
12. Valencia X, Higgins JM, Kiener HP, Lee DM, Podrebarac TA, Dascher CC, Watts GF, Mizoguchi E, Simmons B, Patel DD, Bhan AK, Brenner MB. Cadherin-11 provides specific cellular adhesion between fibroblast-like synoviocytes. J Exp Med. 2004;200(12):1673–9. https://doi.org/10.1084/jem.20041545.
13. Bauer S, Jendro M, Wadle A, Kleber S, Stenner F, Dinser R, Reich A, Faccin E, Godde S, Dinges H, Muller-Ladner U, Renner C. Fibroblast activation protein is expressed by rheumatoid myofibroblast-like synoviocytes. Arthritis Res Ther. 2006;8(6):R171.
14. Del Rey MJ, Faré R, Izquierdo E, Usategui A, Rodríguez-Fernández JL, Suárez-Fueyo A, Cañete JD, Pablos JL. Clinicopathological correlations of podoplanin (gp38) expression in rheumatoid synovium and its potential contribution to fibroblast platelet crosstalk. PLoS One. 2014;9(6):e99607. https://doi.org/10.1371/journal.pone.0099607.
15. Ekwall A-K, Eisler T, Anderberg C, Jin C, Karlsson N, Brisslert M, Bokarewa M. The tumour-associated glycoprotein podoplanin is expressed in fibroblast-like synoviocytes of the hyperplastic synovial lining layer in rheumatoid arthritis. Arthritis Res Ther. 2011;13(2):R40.
16. MacFadyen JR, Haworth O, Roberston D, Hardie D, Webster M-T, Morris HR, Panico M, Sutton-Smith M, Dell A, van der Geer P, Wienke D, Buckley CD, Isacke CM. Endosialin (TEM1, CD248) is a marker of stromal fibroblasts and is not selectively expressed on tumour endothelium. FEBS Lett. 2005;579(12):2569–75. https://doi.org/10.1016/j.febslet.2005.03.071.

17. Maia M, de Vriese A, Janssens T, Moons M, van Landuyt K, Tavernier J, Lories RJ, Conway EM. CD248 and its cytoplasmic domain: a therapeutic target for arthritis. Arthritis Rheum. 2010;62(12):3595–606. https://doi.org/10.1002/art.27701.
18. Saalbach A, Aneregg U, Bruns M, Schnabel E, Herrmann K, Haustein UF. Novel fibroblast-specific monoclonal antibodies: properties and specificities. J Invest Dermatol. 1996;106(6):1314–9.
19. Saalbach A, Kraft R, Herrmann K, Haustein UF, Anderegg U. The monoclonal antibody AS02 recognizes a protein on human fibroblasts being highly homologous to Thy-1. Arch Dermatol Res. 1998;290(7):360–6.
20. Gitter BD, Labus JM, Lees SL, Scheetz ME. Characteristics of human synovial fibroblast activation by IL-1 beta and TNF alpha. Immunology. 1989;66(2):196–200.
21. Ospelt C, Brentano F, Rengel Y, Stanczyk J, Kolling C, Tak PP, Gay RE, Gay S, Kyburz D. Overexpression of Toll-like receptors 3 and 4 in synovial tissue from patients with early rheumatoid arthritis: Toll-like receptor expression in early and longstanding arthritis. Arthritis Rheum. 2008;58(12):3684–92. https://doi.org/10.1002/art.24140.
22. Okamoto H, Yamamura M, Morita Y, Harada S, Makino H, Ota Z. The synovial expression and serum levels of interleukin-6, interleukin-11, leukemia inhibitory factor, and oncostatin M in rheumatoid arthritis. Arthritis Rheum. 1997;40(6):1096–105. https://doi.org/10.1002/1529-0131(199706)40:6<1096::AID-ART13>3.0.CO;2-D.
23. Hwang SY, Kim JY, Kim KW, Park MK, Moon Y, Kim WU, Kim HY. IL-17 induces production of IL-6 and IL-8 in rheumatoid arthritis synovial fibroblasts via NF-kappaB- and PI3-kinase/Akt-dependent pathways. Arthritis Res Ther. 2004;6(2):R120–8. https://doi.org/10.1186/ar1038.
24. Parsonage G, Filer A, Bik M, Hardie D, Lax S, Howlett K, Church LD, Raza K, Wong SH, Trebilcock E, Scheel-Toellner D, Salmon M, Lord JM, Buckley CD. Prolonged, granulocyte-macrophage colony-stimulating factor-dependent, neutrophil survival following rheumatoid synovial fibroblast activation by IL-17 and TNFalpha. Arthritis Res Ther. 2008;10(2):R47. https://doi.org/10.1186/ar2406.
25. Zucali JR, Dinarello CA, Oblon DJ, Gross MA, Anderson L, Weiner RS. Interleukin 1 stimulates fibroblasts to produce granulocyte-macrophage colony-stimulating activity and prostaglandin E2. J Clin Investig. 1986;77(6):1857–63.
26. Donlin LT, Jayatilleke A, Giannopoulou EG, Kalliolias GD, Ivashkiv LB. Modulation of TNF-induced macrophage polarization by synovial fibroblasts. J Immunol. 2014;193(5):2373–83. https://doi.org/10.4049/jimmunol.1400486.
27. Chomarat P, Rissoan MC, Pin JJ, Banchereau J, Miossec P. Contribution of IL-1, CD14, and CD13 in the increased IL-6 production induced by in vitro monocyte-synoviocyte interactions. J Immunol. 1995;155(7):3645–52.
28. Bradfield PF, Amft N, Vernon-Wilson E, Exley AE, Parsonage G, Rainger GE, Nash GB, Thomas AM, Simmons DL, Salmon M, Buckley CD. Rheumatoid fibroblast-like synoviocytes overexpress the chemokine stromal cell-derived factor 1 (CXCL12), which supports distinct patterns and rates of CD4+ and CD8+ T cell migration within synovial tissue. Arthritis Rheum. 2003;48(9):2472–82. https://doi.org/10.1002/art.11219.
29. Burger JA, Zvaifler NJ, Tsukada N, Firestein GS, Kipps TJ. Fibroblast-like synoviocytes support B-cell pseudoemperipolesis via a stromal cell-derived factor-1- and CD106 (VCAM-1)-dependent mechanism. J Clin Invest. 2001;107(3):305–15. https://doi.org/10.1172/jci11092.
30. Scaife S, Brown R, Kellie S, Filer A, Martin S, Thomas AMC, Bradfield PF, Amft N, Salmon M, Buckley CD. Detection of differentially expressed genes in synovial fibroblasts by restriction fragment differential display. Rheumatology. 2004;43(11):1346–52. https://doi.org/10.1093/rheumatology/keh347.
31. Pierer M, Rethage J, Seibl R, Lauener R, Brentano F, Wagner U, Hantzschel H, Michel BA, Gay RE, Gay S, Kyburz D. Chemokine secretion of rheumatoid arthritis synovial fibroblasts stimulated by Toll-like receptor 2 ligands. J Immunol. 2004;172(2):1256–65. https://doi.org/10.4049/jimmunol.172.2.1256.

32. Kim KW, Cho ML, Kim HR, Ju JH, Park MK, Oh HJ, Kim JS, Park SH, Lee SH, Kim HY. Up-regulation of stromal cell-derived factor 1 (CXCL12) production in rheumatoid synovial fibroblasts through interactions with T lymphocytes: role of interleukin-17 and CD40L-CD40 interaction. Arthritis Rheum. 2007;56(4):1076–86. https://doi.org/10.1002/art.22439.

33. Hayashida K, Nanki T, Girschick H, Yavuz S, Ochi T, Lipsky P. Synovial stromal cells from rheumatoid arthritis patients attract monocytes by producing MCP-1 and IL-8. Arthritis Res. 2001;3(2):1–9. https://doi.org/10.1186/ar149.

34. Chan A, Filer A, Parsonage G, Kollnberger S, Gundle R, Buckley CD, Bowness P. Mediation of the proinflammatory cytokine response in rheumatoid arthritis and spondylarthritis by interactions between fibroblast-like synoviocytes and natural killer cells. Arthritis Rheum. 2008;58(3):707–17. https://doi.org/10.1002/art.23264.

35. Benito-Miguel M, García-Carmona Y, Balsa A, Bautista-Caro M-B, Arroyo-Villa I, Cobo-Ibáñez T, Bonilla-Hernán MG, de Ayala CP, Sánchez-Mateos P, Martín-Mola E, Miranda-Carús M-E. IL-15 expression on RA synovial fibroblasts promotes B cell survival. PLoS One. 2012;7(7):e40620. https://doi.org/10.1371/journal.pone.0040620.

36. Bombardieri M, Kam NW, Brentano F, Choi K, Filer A, Kyburz D, McInnes IB, Gay S, Buckley C, Pitzalis C. A BAFF/APRIL-dependent TLR3-stimulated pathway enhances the capacity of rheumatoid synovial fibroblasts to induce AID expression and Ig class-switching in B cells. Ann Rheum Dis. 2011;70(10):1857–65. https://doi.org/10.1136/ard.2011.150219.

37. Filer A, Parsonage G, Smith E, Osborne C, Thomas AM, Curnow SJ, Rainger GE, Raza K, Nash GB, Lord J, Salmon M, Buckley CD. Differential survival of leukocyte subsets mediated by synovial, bone marrow, and skin fibroblasts: site-specific versus activation-dependent survival of T cells and neutrophils. Arthritis Rheum. 2006;54(7):2096–108. https://doi.org/10.1002/art.21930.

38. Muller-Ladner U, Kriegsmann J, Franklin BN, Matsumoto S, Geiler T, Gay RE, Gay S. Synovial fibroblasts of patients with rheumatoid arthritis attach to and invade normal human cartilage when engrafted into SCID mice. Am J Pathol. 1996;149(5):1607–15.

39. Unemori EN, Hibbs MS, Amento EP. Constitutive expression of a 92-kD gelatinase (type V collagenase) by rheumatoid synovial fibroblasts and its induction in normal human fibroblasts by inflammatory cytokines. J Clin Invest. 1991;88(5):1656–62. https://doi.org/10.1172/jci115480.

40. Pap T, Aupperle KR, Gay S, Firestein GS, Gay RE. Invasiveness of synovial fibroblasts is regulated by p53 in the SCID mouse in vivo model of cartilage invasion. Arthritis Rheum. 2001;44(3):676–81. https://doi.org/10.1002/1529-0131(200103)44:3<676::aid-anr117>3.0.co;2-6.

41. Lefevre S, Knedla A, Tennie C, Kampmann A, Wunrau C, Dinser R, Korb A, Schnaker EM, Tarner IH, Robbins PD, Evans CH, Sturz H, Steinmeyer J, Gay S, Scholmerich J, Pap T, Muller-Ladner U, Neumann E. Synovial fibroblasts spread rheumatoid arthritis to unaffected joints. Nat Med. 2009;15(12):1414–20. https://doi.org/10.1038/nm.2050.

42. Huybrechts-Godin G, Hauser P, Vaes G. Macrophage-fibroblast interactions in collagenase production and cartilage degradation. Biochem J. 1979;184(3):643–50.

43. Janusz MJ, Hare M. Cartilage degradation by cocultures of transformed macrophage and fibroblast cell lines. A model of metalloproteinase-mediated connective tissue degradation. J Immunol. 1993;150(5):1922–31.

44. Scott BB, Weisbrot LM, Greenwood JD, Bogoch ER, Paige CJ, Keystone EC. Rheumatoid arthritis synovial fibroblast and U937 macrophage/monocyte cell line interaction in cartilage degradation. Arthritis Rheum. 1997;40(3):490–8.

45. Portela A, Esteller M. Epigenetic modifications and human disease. Nat Biotechnol. 2010;28(10):1057–68. https://doi.org/10.1038/nbt.1685.

46. Wu SC, Zhang Y. Active DNA demethylation: many roads lead to Rome. Nat Rev Mol Cell Biol. 2010;11(9):607–20. http://www.nature.com/nrm/journal/v11/n9/suppinfo/nrm2950_S1.html

47. Huber L, Brock M, Hemmatazad H, Giger O, Moritz F, Trenkmann M, Distler J, Gay R, Kolling C, Moch H, Michel B, Gay S, Distler O, Jungel A. Histone deacetylase/acetylase activity in total synovial tissue derived from rheumatoid arthritis and osteoarthritis patients. Arthritis Rheum. 2007;56:1087–93.

48. Kawabata T, Nishida K, Takasugi K, Ogawa H, Sada K, Kadota Y, Inagaki J, Hirohata S, Ninomiya Y, Makino H. Increased activity and expression of histone deacetylase 1 in relation to tumor necrosis factor-alpha in synovial tissue of rheumatoid arthritis. Arthritis Res Ther. 2010;12:R133.

49. Wada TT, Araki Y, Sato K, Aizaki Y, Yokota K, Kim YT, Oda H, Kurokawa R, Mimura T. Aberrant histone acetylation contributes to elevated interleukin-6 production in rheumatoid arthritis synovial fibroblasts. Biochem Biophys Res Commun. 2014;444(4):682–6. https://doi.org/10.1016/j.bbrc.2014.01.195.

50. Grabiec A, Korchynskyi O, Tak P, Reedquist K. Histone deacetylase inhibitors suppress rheumatoid arthritis fibroblast-like synoviocyte and macrophage IL-6 production by accelerating mRNA decay. Ann Rheum Dis. 2012;71:424–31.

51. Karouzakis E, Gay R, Michel B, Gay S, Neidhart M. DNA hypomethylation in rheumatoid arthritis synovial fibroblasts. Arthritis Rheum. 2009;60:3613–22.

52. de la Rica L, Urquiza JM, Gómez-Cabrero D, Islam ABMMK, López-Bigas N, Tegnér J, Toes REM, Ballestar E. Identification of novel markers in rheumatoid arthritis through integrated analysis of DNA methylation and microRNA expression. J Autoimmun. 2013;41(0):6–16. https://doi.org/10.1016/j.jaut.2012.12.005.

53. He L, Hannon GJ. MicroRNAs: small RNAs with a big role in gene regulation. Nat Rev Genet. 2004;5(7):522–31.

54. Long L, Yu P, Liu Y, Wang S, Li R, Shi J, Zhang X, Li Y, Sun X, Zhou B, Cui L, Li Z. Upregulated microRNA-155 expression in peripheral blood mononuclear cells and fibroblast-like synoviocytes in rheumatoid arthritis. Clin Dev Immunol. 2013;2013:296139. https://doi.org/10.1155/2013/296139.

55. Stanczyk J, Pedrioli DML, Brentano F, Sanchez-Pernaute O, Kolling C, Gay RE, Detmar M, Gay S, Kyburz D. Altered expression of MicroRNA in synovial fibroblasts and synovial tissue in rheumatoid arthritis. Arthritis Rheum. 2008;58(4):1001–9. https://doi.org/10.1002/art.23386.

56. Lin J, Huo R, Xiao L, Zhu X, Xie J, Sun S, He Y, Zhang J, Sun Y, Zhou Z, Wu P, Shen B, Li D, Li N. A novel p53/microRNA-22/Cyr61 axis in synovial cells regulates inflammation in rheumatoid arthritis. Arthritis Rheum. 2014;66(1):49–59. https://doi.org/10.1002/art.38142.

57. Bar-Shavit Z. The osteoclast: a multinucleated, hematopoietic-origin, bone-resorbing osteoimmune cell. J Cell Biochem. 2007;102(5):1130–9. https://doi.org/10.1002/jcb.21553.

58. Sims NA, Gooi JH. Bone remodeling: multiple cellular interactions required for coupling of bone formation and resorption. Semin Cell Dev Biol. 2008;19(5):444–51. https://doi.org/10.1016/j.semcdb.2008.07.016.

59. Goldring SR. Periarticular bone changes in rheumatoid arthritis: pathophysiological implications and clinical utility. Ann Rheum Dis. 2009;68(3):297–9. https://doi.org/10.1136/ard.2008.099408.

60. Klareskog L, van der Heijde D, de Jager JP, Gough A, Kalden J, Malaise M, Martin Mola E, Pavelka K, Sany J, Settas L, Wajdula J, Pedersen R, Fatenejad S, Sanda M. Therapeutic effect of the combination of etanercept and methotrexate compared with each treatment alone in patients with rheumatoid arthritis: double-blind randomised controlled trial. Lancet. 2004;363(9410):675–81. https://doi.org/10.1016/s0140-6736(04)15640-7.

61. Claes L, Recknagel S, Ignatius A. Fracture healing under healthy and inflammatory conditions. Nat Rev Rheumatol. 2012;8(3):133–43. https://doi.org/10.1038/nrrheum.2012.1.

62. Long F. Building strong bones: molecular regulation of the osteoblast lineage. Nat Rev Mol Cell Biol. 2012;13(1):27–38. https://doi.org/10.1038/nrm3254.

63. Walsh NC, Reinwald S, Manning CA, Condon KW, Iwata K, Burr DB, Gravallese EM. Osteoblast function is compromised at sites of focal bone erosion in inflammatory arthritis. J Bone Miner Res. 2009;24(9):1572–85. https://doi.org/10.1359/jbmr.090320.
64. Gilbert L, He X, Farmer P, Boden S, Kozlowski M, Rubin J, Nanes MS. Inhibition of osteoblast differentiation by tumor necrosis factor-alpha. Endocrinology. 2000;141(11):3956–64. https://doi.org/10.1210/endo.141.11.7739.
65. Gilbert L, He X, Farmer P, Rubin J, Drissi H, van Wijnen AJ, Lian JB, Stein GS, Nanes MS. Expression of the osteoblast differentiation factor RUNX2 (Cbfa1/AML3/Pebp2alpha A) is inhibited by tumor necrosis factor-alpha. J Biol Chem. 2002;277(4):2695–701. https://doi.org/10.1074/jbc.M106339200.
66. Bertolini DR, Nedwin GE, Bringman TS, Smith DD, Mundy GR. Stimulation of bone resorption and inhibition of bone formation in vitro by human tumour necrosis factors. Nature. 1986;319(6053):516–8. https://doi.org/10.1038/319516a0.
67. Centrella M, McCarthy TL, Canalis E. Tumor necrosis factor-alpha inhibits collagen synthesis and alkaline phosphatase activity independently of its effect on deoxyribonucleic acid synthesis in osteoblast-enriched bone cell cultures. Endocrinology. 1988;123(3):1442–8. https://doi.org/10.1210/endo-123-3-1442.
68. Li YP, Stashenko P. Proinflammatory cytokines tumor necrosis factor-alpha and IL-6, but not IL-1, down-regulate the osteocalcin gene promoter. J Immunol. 1992;148(3):788–94.
69. Panagakos FS, Fernandez C, Kumar S. Ultrastructural analysis of mineralized matrix from human osteoblastic cells: effect of tumor necrosis factor-alpha. Mol Cell Biochem. 1996;158(1):81–9.
70. Liu XH, Kirschenbaum A, Yao S, Levine AC. Cross-talk between the interleukin-6 and prostaglandin E(2) signaling systems results in enhancement of osteoclastogenesis through effects on the osteoprotegerin/receptor activator of nuclear factor-{kappa}B (RANK) ligand/RANK system. Endocrinology. 2005;146(4):1991–8. https://doi.org/10.1210/en.2004-1167.
71. Musacchio E, Valvason C, Botsios C, Ostuni F, Furlan A, Ramonda R, Modesti V, Sartori L, Punzi L. The tumor necrosis factor-{alpha}-blocking agent infliximab inhibits interleukin 1beta (IL-1beta) and IL-6 gene expression in human osteoblastic cells. J Rheumatol. 2009;36(8):1575–9. https://doi.org/10.3899/jrheum.081321.
72. Hengartner NE, Fiedler J, Ignatius A, Brenner RE. IL-1beta inhibits human osteoblast migration. Mol Med. 2013;19:36–42. https://doi.org/10.2119/molmed.2012.00058.
73. Nguyen L, Dewhirst FE, Hauschka PV, Stashenko P. Interleukin-1 beta stimulates bone resorption and inhibits bone formation in vivo. Lymphokine Cytokine Res. 1991;10(1–2):15–21.
74. Stashenko P, Dewhirst FE, Rooney ML, Desjardins LA, Heeley JD. Interleukin-1 beta is a potent inhibitor of bone formation in vitro. J Bone Miner Res. 1987;2(6):559–65. https://doi.org/10.1002/jbmr.5650020612.
75. Almeida M, Han L, Martin-Millan M, O'Brien CA, Manolagas SC. Oxidative stress antagonizes Wnt signaling in osteoblast precursors by diverting beta-catenin from T cell factor- to forkhead box O-mediated transcription. J Biol Chem. 2007;282(37):27298–305. https://doi.org/10.1074/jbc.M702811200.
76. Brandao-Burch A, Utting JC, Orriss IR, Arnett TR. Acidosis inhibits bone formation by osteoblasts in vitro by preventing mineralization. Calcif Tissue Int. 2005;77(3):167–74. https://doi.org/10.1007/s00223-004-0285-8.
77. Colla S, Zhan F, Xiong W, Wu X, Xu H, Stephens O, Yaccoby S, Epstein J, Barlogie B, Shaughnessy JD Jr. The oxidative stress response regulates DKK1 expression through the JNK signaling cascade in multiple myeloma plasma cells. Blood. 2007;109(10):4470–7. https://doi.org/10.1182/blood-2006-11-056747.
78. Kim KW, Cho ML, Lee SH, Oh HJ, Kang CM, Ju JH, Min SY, Cho YG, Park SH, Kim HY. Human rheumatoid synovial fibroblasts promote osteoclastogenic activity by activating RANKL via TLR-2 and TLR-4 activation. Immunol Lett. 2007;110(1):54–64. https://doi.org/10.1016/j.imlet.2007.03.004.

79. Kim KW, Cho ML, Oh HJ, Kim HR, Kang CM, Heo YM, Lee SH, Kim HY. TLR-3 enhances osteoclastogenesis through upregulation of RANKL expression from fibroblast-like synoviocytes in patients with rheumatoid arthritis. Immunol Lett. 2009;124(1):9–17. https://doi.org/10.1016/j.imlet.2009.02.006.

80. Matzelle MM, Gallant MA, Condon KW, Walsh NC, Manning CA, Stein GS, Lian JB, Burr DB, Gravallese EM. Resolution of inflammation induces osteoblast function and regulates the Wnt signaling pathway. Arthritis Rheum. 2012;64(5):1540–50. https://doi.org/10.1002/art.33504.

81. Diarra D, Stolina M, Polzer K, Zwerina J, Ominsky MS, Dwyer D, Korb A, Smolen J, Hoffmann M, Scheinecker C, van der Heide D, Landewe R, Lacey D, Richards WG, Schett G. Dickkopf-1 is a master regulator of joint remodeling. Nat Med. 2007;13(2):156–63. https://doi.org/10.1038/nm1538.

82. Uderhardt S, Diarra D, Katzenbeisser J, David JP, Zwerina J, Richards W, Kronke G, Schett G. Blockade of Dickkopf (DKK)-1 induces fusion of sacroiliac joints. Ann Rheum Dis. 2010;69(3):592–7. https://doi.org/10.1136/ard.2008.102046.

83. Chen XX, Baum W, Dwyer D, Stock M, Schwabe K, Ke HZ, Stolina M, Schett G, Bozec A. Sclerostin inhibition reverses systemic, periarticular and local bone loss in arthritis. Ann Rheum Dis. 2013;72(10):1732–6. https://doi.org/10.1136/annrheumdis-2013-203345.

84. Jungel A, Baresova V, Ospelt C, Simmen B, Michel B, Gay R, Gay S, Seemayer C, Neidhart M. Trichostatin A sensitises rheumatoid arthritis synovial fibroblasts for TRAIL-induced apoptosis. Ann Rheum Dis. 2006;65:910–2.

85. Nasu Y, Nishida K, Miyazawa S, Komiyama T, Kadota Y, Abe N, Yoshida A, Hirohata S, Ohtsuka A, Ozaki T. Trichostatin A, a histone deacetylase inhibitor, suppresses synovial inflammation and subsequent cartilage destruction in a collagen antibody-induced arthritis mouse model. Osteoarthr Cartil. 2008;16:723–32.

86. Vojinovic J, Damjanov N, D'Urzo C, Furlan A, Susic G, Pasic S, Iagaru N, Stefan M, Dinarello CA. Safety and efficacy of an oral histone deacetylase inhibitor in systemic-onset juvenile idiopathic arthritis. Arthritis Rheum. 2011;63(5):1452–8. https://doi.org/10.1002/art.30238.

87. Yang C-R, Shih K-S, Liou J-P, Wu Y-W, Hsieh IN, Lee H-Y, Lin T-C, Wang J-H. Denbinobin upregulates miR-146a expression and attenuates IL-1β-induced upregulation of ICAM-1 and VCAM-1 expressions in osteoarthritis fibroblast-like synoviocytes. J Mol Med. 2014:1–12. https://doi.org/10.1007/s00109-014-1192-8.

88. Yao Y, Jia T, Pan Y, Gou H, Li Y, Sun Y, Zhang R, Zhang K, Lin G, Xie J, Li J, Wang L. Using a novel microRNA delivery system to inhibit osteoclastogenesis. Int J Mol Sci. 2015;16(4):8337–50.

Chapter 4
Molecular and Cellular Requirements for the Assembly of Tertiary Lymphoid Structures

C. G. Mueller, S. Nayar, J. Campos, and F. Barone

Abstract At sites of chronic inflammation, recruited immune cells form structures that resemble secondary lymphoid organs (SLOs). Those are characterized by segregated areas of prevalent T- or B-cell aggregation, differentiation of high endothelial venules (HEVs) and local activation of resident stromal cells. B-cell proliferation and affinity maturation towards locally displayed autoantigens have been demonstrated at those sites, known as tertiary lymphoid structures (TLSs). TLS formation has been associated with local disease persistence and progression as well as increased systemic manifestations. While bearing a similar histological structure to SLO, the signals that regulate TLS and SLO formation can diverge, and a series of pro-inflammatory cytokines has been ascribed as responsible for TLS formation at different anatomical sites. Here we review the structural elements as well as the signals responsible for TLS aggregation, aiming to provide an overview to this complex immunological phenomenon.

Keywords Tertiary lymphoid structures · TNF · Lymphotoxin · RANKL
Endothelial and stromal cells · CXCL13 · CCL21 · Sjögren's syndrome

C. G. Mueller
CNRS UPR 3572, Laboratory of Immunopathology and Therapeutic Chemistry/
Laboratory of Excellence MEDALIS, Institut de Biologie Moléculaire et Cellulaire,
Université de Strasbourg, Strasbourg, France

S. Nayar · J. Campos · F. Barone (✉)
Rheumatology Research Group, Institute of Inflammation and Ageing (IIA),
University of Birmingham, Birmingham, UK
e-mail: f.barone@bham.ac.uk

© Springer International Publishing AG, part of Springer Nature 2018
B. M.J Owens, M. A. Lakins (eds.), *Stromal Immunology*, Advances in Experimental
Medicine and Biology 1060, https://doi.org/10.1007/978-3-319-78127-3_4

4.1 Introduction

4.1.1 Definition of Tertiary Lymphoid Organs

Tertiary lymphoid structures (TLSs), also named ectopic lymphoid structures, are best defined as the organoid assembly of cells of the adaptive immune system (B and T lymphocytes) in non-immune tissue. They comprise one or more follicles of B cells that may cluster around fibroblastic stromal cells, usually referred to as follicular dendritic cells (FDCs) [1]. TLSs are also characterized by T cells that, with interdigitating dendritic cells (DCs), collect around fibroblastic reticular cells (FRCs) [2]. These structures are vascularized, and the blood endothelial cells express high levels of chemokines and integrins to actively recruit leucocytes. Endothelial cells and fibroblasts express cell adhesion factors such as MAdCAM-1 (mucosal vascular addressin cell adhesion molecule 1), VCAM-1 (vascular cell adhesion molecule 1) or ICAM-1 (intercellular adhesion molecule 1). HEVs can also express PNAd (peripheral node addressin), the ligand for L selectin, to maximize the recruitment of lymphocytes from the bloodstream [3]. In addition, lymphatic endothelial cells can also be found at sites of TLS development; however, on the contrary to what is seen in lymph node architecture, connective tissue encapsulating TLSs is a rare finding.

From a purely conceptual point of view, TLSs develop when the forces that recruit and retain leucocytes exceed those expulsing them from the tissue. This can occur either through an overactive recruitment via blood endothelial cells or by a diminished lymphatic endothelial cell-regulated cell output. Nonetheless, lymphedema is unlikely to be sufficient to create TLSs when the retaining force is underdeveloped. Therefore, a coordinated interplay between entry-retention-exit is probably required for TLS formation. Moreover, lymphocytes retained in the tissue rarely achieve the level of organization sufficient to form a TLS, thus suggesting that an active process of recruitment and organization is required for TLS formation [4].

TLSs arise in tissues whose main function is other than the generation of immune cells or the initiation of an adaptive immune response. This excludes the bone marrow, thymus (primary lymphoid organs) or spleen, lymph nodes and Peyer's patches (SLOs). The kidney, heart, pancreas, synovium, etc. are regarded as non-immune organs. However, such classification is not as clear for organs like the intestine or the liver. One of the functions of the intestine is to protect the body against potential pathogenic microflora, and Peyer's patches, cryptopatches and isolated lymphoid follicles arise as part of this function in response to normal living conditions. Therefore, these structures should not be regarded as TLSs. On the contrary, the liver fulfils a haematopoietic function in the embryo [5] that then fades into negligence in the adult. The function of the adult liver is no longer to provide haematopoietic cells but to run the body's chemical powerhouse. Therefore, an assembly of organized lymphocytes in the adult liver should be regarded as TLSs. As for the lung or the salivary glands, these organs either have an efficient innate immune system or can rely on efficient drainage to SLOs to combat pathogens without the need of TLSs. Therefore, in non-pathogenic condition, salivary glands and lungs normally do not comprise lymphoid structures, and any organoid immune cell assembly arising there should be considered as TLSs (Fig. 4.1) [6, 7].

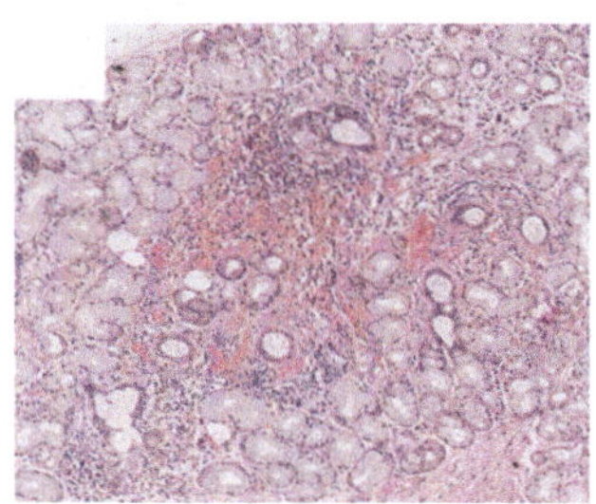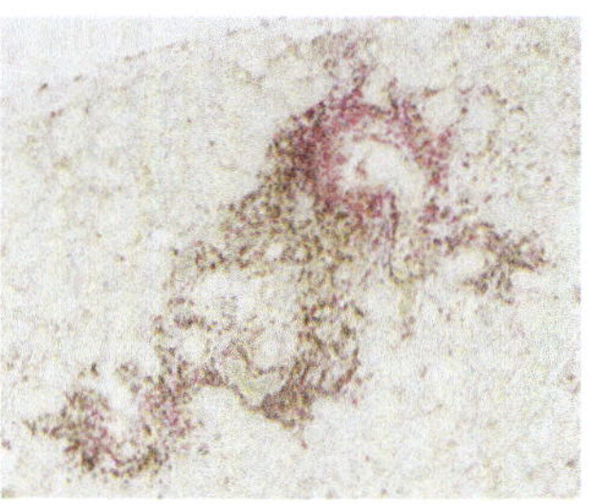

Fig. 4.1 a. and b. Microphotographs illustrating TLS formation in the salivary glands of a patient with Sjogren's Syndrome. Sequential section showing T/B cell segregation (CD3, brown and CD20 in pink in b)

4.1.2 Function of TLSs

An important function associated with TLS formation is local production of antibodies. TLSs do so by providing T- and B-cell survival factors, IL-7 and B-cell activating factor (BAFF), to locally recruited lymphocytes and favouring the interaction between these cell types in a confined environment [8]. Local B-cell activation has been demonstrated by expression of AICDA (the enzyme responsible for class switch recombination and somatic hypermutation) [9] and active proliferation in ectopic germinal centre-like structures. Local differentiation of autoreactive plasma cells has also been shown [10].

Ectopic expression of lymphoid or homeostatic chemokines, known to regulate naïve and central memory T margination, CCL21 and CCL19, and B-cell organization in follicles and germinal centres, CXCL13 and CXCL12, are also found in TLSs [11, 12].

Transient formation of TLSs can occur in physiological settings in response to pathogens, and, in these cases [4, 13, 14], TLSs are believed to contribute to the generation of antigen-specific B cells to fight local infection [4, 15].

During chronic inflammation, for example, in the salivary glands of patients with Sjögren's syndrome (SS) and in the synovium of rheumatoid arthritis (RA), the presence of TLSs is strongly associated with disease progression rather than resolution. TLS formation correlates with autoantibody serum levels and disease severity in several autoimmune diseases and in animal models of diabetes and SS [7, 16, 17]. In patients with SS, the presence of TLSs is associated with higher levels of circulating autoantibodies and systemic manifestations [10, 18–20]. TLSs that form during RA in the subchondral bone contribute to osteoclast activation and tissue damage. In addition, the levels of CXCL13, a chemokine canonically associated with TLS formation, correlate with disease severity in RA, and the persistence of subclinical synovitis is detected by ultrasound [21, 22].

TLS-associated B-cell activation is a recognized mechanism of lymphoma progression in the salivary glands of patients with SS and in the gastric mucosa of patients with *Helicobacter pylori* gastritis [23–26]. In SS the identification of ectopic TLSs with fully formed germinal centres (GC) within the minor salivary glands is currently used as histological biomarker and prognostic tool for lymphoma development [20].

In contrast, TLSs that form in the vascular adventitia during atherosclerosis inhibit disease progression through a mechanism that involves lymphotoxin-β receptor (LTβR) signalling [27]. Similar observations support an immunosuppressive role for TLSs in solid tumours [28], thus raising the intriguing possibility that TLS function is contextual and their pro- or anti-inflammatory properties are tissue and disease specific.

4.1.3 Spontaneous Versus Induced TLS

TLS can spontaneously arise under conditions of chronic inflammation caused by the persistence of inflammatory signals, self-antigen during autoimmunity or recurrent infections.

In the genetically predisposed nonobese diabetic (NOD) mouse strain, TLSs form in the pancreas, increase in size and acquire highly structural parameters as the disease progresses from peri- to intrainsulitis [16]. Mouse models develop spontaneously TLSs in the adventitial aorta (ATLSs) [29], in the central nervous system (CNS) [30] and in the gastrointestinal tract (stomach) [31, 32]. In humans, TLSs are associated with RA [12, 33] and SS [2].

TLSs can also be generated artificially. The overexpression of chemokines or other organizing molecules and the administration of inflammatory substances or pathogens can lead to the organoid assembly of T or B cells [34]. The site of assembly can be chosen with the appropriate tissue-specific promoter (i.e. the rat insulin promoter of the liver ovalbumin promoter) [35] or by the necessary technique to introduce exogenous material [7, 8].

These two types of TLS differ in two ways, immune activity and persistence. An immune activity of spontaneous TLS is always present but can be considered as low resulting in long-term chronicity. In contrast, that of induced TLSs is variable, depending on the type of stimulus employed to trigger them. For instance, the organized recruitment of immune cells by overexpression of chemokines [36] would likely result in a weak activation, whereas introducing high concentrations of pathogens or pathogen-derived product would lead to an overly active immune response [4]. However, the distinction between spontaneous and induced TLSs based on immune activity is ambiguous, since there is no clear measure to grade immune activity and its assessment is highly subjective. In addition, it is possible that immune activity evolves, for instance, the gradual lymphocyte accumulation can disturb organ function and then result in overt inflammation.

A more useful way to distinguish between spontaneous and induced TLSs is maintenance. The elimination of the signal that induced the organoid immune cell assembly should also lead to its disappearance [37]. This can be tested using a promoter for gene overexpression whose activity is controllable or using pathogen or pathogen-derived products with a short half-life. In the case of the spontaneously arising TLS, the signal that triggered its formation is complex, redundant and partially not understood, hence difficult to eliminate, which leads to persistence. This aspect raises much interest in the scientific and medical community, since the identification of those inductive signals would allow their medical targeting to resolve TLSs. From this point of view, also induced TLSs have great value since they allow the identification of those molecules capable of inducing TLSs (and involved in their development and maturation) and provide a rapid study system to test therapeutic efficacy.

4.2 Molecular Cues for TLS Formation

The studies of experimentally induced TLS have largely contributed to our understanding of the mechanisms that lead to TLSs, and they have co-evolved with the dissection of the molecular programming that underlies the normal development of SLOs. Indeed, the finding that the same molecular programmes that control SLO development also induce TLS formation led to the notion that TLSs greatly resemble SLOs in structure and function. However, it is still premature to conclude that these concepts can be generally applied to prevent their formation or resolve existing TLSs.

4.2.1 Chemokines

CXCL13 is expressed by fibroblastic stromal cells and is a key chemokine for B cells and lymphoid tissue inducer (LTi) cells. Mice deficient in CXCL13 lack all lymph nodes except facial, cervical and mesenteric lymph nodes [38]. Its overexpression by the rat insulin promoter (RIP), active in the pancreas and kidney, a popular model system for induced TLS formation [39], leads to TLS formation characterized by segregated B-/T-cell zones, the presence of conventional DCs and a dense network of stromal cells and HEV-type blood vessels [35]. Increased expression of CXCL13 and B-cell infiltration was also described in the central nervous tissue of mice in experimental autoimmune encephalomyelitis [30]. Among other chemokines (notably CXCL10), CXCL13 has been found in the spontaneous mouse model of autoimmune gastritis [31]. The spontaneous TLSs in the pancreas of diabetic NOD mice show a local upregulation of CXCL13, CXCL12 and CCL19, concomitant with FDC formation and B-cell activation, while the mRNA expression levels of CXCR5, CCR7 and CXCR4 increase less markedly [16]. In a mouse model of ATLSs, aorta smooth muscle actin-expressing cells synthesized CXCL13 and CCL21 [29]. CXCL13, CCL21 and CXCL12 were found in Sjögren's syndrome tissue but varied according to the different clinical stages [17, 23, 40]. Hashimoto's thyroiditis [41] ectopic lymphoid tissue formation in the thyroid gland shows the presence of CXCL13 mRNA that correlates with CXCR5 mRNA levels and the number of focal lymphocytic infiltrates and germinal centres [41]. CXCL13 is present in rheumatoid arthritis and plays a predominant role over CCL21 in lymphoid foci formation [33]. Indeed, there is some evidence suggesting that CXCL13 and LTb expression might predict the development of ectopic germinal centres in patients with RA and SS [3, 17, 42, 43].

CXCL12 (or stromal cell-derived factor 1, SDF1) is critical in bone marrow haematopoiesis and B-cell development, where it is expressed by bone marrow stromal cells [44]. It is displayed by HEVs in SLOs and acts as an important B cell recruiting chemokine, while T cells are mostly unresponsive [45]. Therefore, and not surprisingly, RIP (rat insulin promoter)-CXCL12 transgenic mice presented with small infiltrates comprising few T cells but enriched in DCs, B cells and plasma cells [46]. Significant upregulation of CXCL12 is observed in TLS associated with lymphoma development in the salivary glands of patients with SS [23].

CCL19 and CCL21, expressed by endothelial cells and some stromal cells, are ligands for CCR7 carried by T cells, DCs and LTi cells. A critical role for CCR7 and CCL19/CCL21 in T-cell homing was shown by *plt* mice that lack the CCL19 gene and the CCL21-ser expressed by lymphatic vessels of the lymphoid tissue. In the RIP overexpression model, CCL21 appears more effective than CCL19 in forming ectopic lymphoid structures [46, 47]; however, even with CCL21 overexpression, a distinctive B-cell follicle fails to form [46]. CCL19 and CCL21 have both been detected in ectopic infiltrates of RA and SS [12, 33].

Ectopic expression of CCL21 in the thyroid gland was sufficient to induce TLS formation that resembled the structures seen in Hashimoto's thyroiditis and Graves' disease [48]. As neither of the two CCL21 transgenic models [46, 48] presented evidence for CD35+ FDCs or CXCL13+ stromal cells, these studies suggested that CCL19 or CCL21 overexpression alone is not sufficient to induce complete lymphoid tissue neogenesis.

4.2.2 TNFSF Members

The TNFSF (tumour necrosis factor superfamily) members TNFα, lymphotoxin (LT) α and β and their signalling receptors TNFRI/II and LTβR were promptly suggested to promote the formation of TLSs when their critical role in SLO development emerged. Seminal work by Ruddle and her group showed that ectopic expression of TNFα or LTα, but not LTβ, under the control of rat insulin promoter led to formation of TLSs [49, 50]. The strongest effect was seen when LTα and LTβ were co-expressed, resulting in an invasive leukocyte accumulation of the pancreatic islets and significantly larger TLSs than in LTα transgenic mice [49]. Moreover, the HEVs were characterized by luminal PNAd expression, thus providing the molecular mechanisms for naïve T-cell and B-cell recruitment [49]. Of the two TNFRs, TNFRI, the principal mediator of lymphoid tissue organogenesis and germinal centre reaction [51], plays the major role in mediating LTα-induced pancreatic TLS [52]. Investigators have demonstrated that LTα expression in tumour cells leads to the formation of intratumoural lymphoid tissue able to sustain an efficient immune response [53]. Activation of LTβR and TNFRI was implicated in aortic

TLS, where interruption of the LTβR signalling suppressed CXCL13 and CCL21 expression, reduced HEVs formation and disrupted TLS structure and maintenance [29, 54]. In NOD mice, pancreatic TLSs show local upregulation of LTαβ and LIGHT, an alternative LTβR ligand [55, 56]. In contrast, inducible bronchus-associated lymphoid tissues (iBALT), tear duct-associated lymphoid tissues and nasopharynx-associated lymphoid tissues (NALTs) appear to develop independently of LTαβ and LTβR [7, 57–59]. However, LT signalling is crucial for maintenance and organization of these structures in the infected lung tissues, and TLSs are disrupted in LTα-deficient mice [7]. The growth of a lymphatic network in this model is dependent on LTβR signalling [60].

While an effect of LTα, alone or with LTβ, appears to be evident, the role of TNFα is conflicting. In some inflammatory diseases, including those with TLS presence, TNFα exhibits anti-inflammatory activity [61]. For instance, insulitis in NOD mice and lupus in New Zealand lupus-prone mice are improved after injection of TNFα [62, 63].

LIGHT is an alternative ligand for LTβR, and its transgenic overexpression drives TLS formation in animal models of melanoma and fibrosarcoma [53, 64]. In the TLSs of NOD mice, there was a local upregulation of LTαβ and LIGHT [16]. Pancreatic LIGHT overexpression in NOD mice exacerbates the disease [56].

BAFF regulates B-cell survival and is highly expressed in the meninges-associated ectopic GC in a mouse model of CNS inflammation [30]. Although different patterns of lymphoid arrangements usually coexist, TLSs harbouring highly organized ectopic lymphoid follicles tend to express significantly higher levels of LTα, CXCL13 and CCL21 than those with diffuse lymphoid infiltrates [12, 33, 65–67]. In fact, the expression levels of CXCL13 and LTβ may be highly predictive of the presence of ectopic germinal centres in synovial biopsies of patients with RA and SS [12, 33, 43, 65–67].

4.2.3 Cytokines

The ubiquitous transgenic co-expression of IL-6 and IL-6R leads to perivascular accumulation of lymphocytes with an important proportion of B cells and mature plasma B cells [68]. Overexpression of IL-5 in the respiratory epithelium also results in development of organized iBALT. However, unlike the models of homeostatic chemokines or TNF-family ligands mentioned above, which do not develop diabetes or thyroiditis despite TLS formation, the IL-5-dependent induction of iBALT leads to epithelial hypertrophy, goblet cell hyperplasia, accumulation of eosinophils in the airway lumen and peribronchial areas and focal collagen deposition, which are all signs of severe lung pathology [69]. Stimulation of T cells with IL-4 or IL-7 induced LTαβ expression, with IL-7 being most potent for CD4$^+$ T cells [46].

The IL-17 gene family plays an important role in the defence against pathogens and has been implicated in various chronic inflammatory contexts. Like TNFRSF members, IL-17 receptor signals via NF-κB. IL-17 T cells are induced by IL-6, TGFβ and IL-23 but inhibited by IL-27. Mice deficient for IL-27 have been shown to develop more severe pathology in a model of induced arthritis and show higher numbers of T_H17 cells in draining lymph nodes and increased IL-17 in serum [70]. Furthermore, in RA patients, the levels of IL-27 are negatively associated with the presence of TLSs, T- and B-cell infiltration in the synovial tissue and levels of IL-17 expression [70].

IL-17 emerged as an important mediator or iBALT induced by lipopolysaccharide [71]. IL-17 induced inflammatory and homeostatic chemokine production in the absence of LTα and LTβ, but the lymphotoxins were required for the differentiation of fibroblastic reticular cells (FRCs), FDCs and HEVs.

Pseudomonas aeruginosa infection induced the pulmonary accumulation of IL-17-producing γδ T cells, triggering CXCL12 production by stromal cells and thus the recruitment of B cells into structures that lack however FDCs [72]. Using a T-cell transgenic animal model of experimental autoimmune encephalomyelitis (EAE) that mimics multiple sclerosis, Peters and colleagues demonstrated that T cells expressing IL-17 cells induce ectopic lymphoid tissues in the central nervous system (CNS) [73]. BALTs induced by *M. tuberculosis* are also dependent on IL-17 and modulated by IL-23 [6].

Studies of human SS have detected IL-22 mRNA in the affected salivary glands [74], and serum levels of IL-22 correlated with clinical manifestations of the disease, including hypergammaglobulinaemia and autoantibody production [75]. In a mouse model of viral-induced SS, inhibition of IL-22 strongly reduces TLS size [8].

IL-7R is expressed by LTi cells, and, together with CXCR5, IL-7 promotes their accumulation in SLOs [38]. IL-7 overexpression led to ectopic lymphoid structures in non-lymphoid tissues, such as in the pancreas or the salivary glands, which were LTα-dependent [76].

4.3 Cellular Requirements for Induced TLSs

4.3.1 Haematopoietic Cells

SLO development depends on the interaction of the haematopoietic LTi cells (CD3-CD4 + IL-7Ra + RANK+) with lymphoid tissue organizers (LTos), cells of mesenchymal origin characterized by expression of VCAM-1, ICAM-1 and MAdCAM-1 [77–79]. In the context of TLS formation, key cell types have been implicated, including LTi, LTo, IL-17-secreting CD4+ T cells and T follicular helper cells (T_{FH}) [80].

The release of IL-7 and RANKL by LTo cells promotes the expression of LTa1B2 by LTi cells, which in turn engages the LTBR on LTo. Such cascade of events leads to homeostatic chemokine release and vascularization by HEVs [80]. In the context of chronic inflammation-associated TLSs, it has been hypothesized that stromal cells found in close relationship with TLSs may acquire LTo-like properties [12, 81, 82].

It was shown that LTi cells express LTαβ in response to IL-7, TNFα and RANKL [83]. They respond to CXCL13 and CXCL12 chemotactic signals and carry integrins that interact with MAdCAM-1 and VCAM-1. LTi cells persist in the adult as innate lymphoid cells of group 3 (ILC3). Hence, the finding that CXCL13, CCL21, CCL19 and CXCL12 are not equal in their ability to promote TLSs may be due, in part, to their differential capacity to attract and maintain LTi/ILC3 cells and promote LTαβ expression.

ILC requirement for TLS formation is debated. ILC3 cell supports isolate lymphoid follicle (ILF) formation, via IL-22 production; but whether those structures can be considered TLSs is argued [4]. Indeed, there is evidence from different animal models that ILCs, including LTi/ILC3 cells, are not essential for the formation of ectopic lymphoid aggregates. In a model of thyroid CCL21 overexpression, TLS formation occurs in the absence of transcription factor Id2 required for LTi/ILC cell maturation [48, 59, 84]. The above-cited models of Th17-stimulated iBALS formation appear to be independent on LTi cells [7]. It is now accepted that in the context of inflammatory conditions, the signals required for TLS maturation can be provided by other cells [59]. B cells and T cells are an alternative source of LTαβ when appropriately stimulated [38, 85]. In the original model of transgenic CXCL13 overexpression under the RIP, TLS formation was dependent on the presence of B cells [35].

DCs contribute to the growth and maintenance of SLOs [84] by providing VEGF and LTαβ to HEVs and by stimulating the expression of CCL21 by FRCs [86, 87]. DC was also shown to play a critical role in the construction of artificial murine lymphoid structures [88]. The presence of DCs is necessary for the maintenance of iBALT (Incducible bronchus-associated lymphoid tissue) in the model of viral pulmonary infection [89]. In the LPS-induction model of lung iBALT (Incducible bronchus-associated lymphoid tissue), CD11c + DCs are necessary for the maintenance of the ectopic lymphoid structures [89]. Myeloid CD68⁺ cells are also a source of chemokines such as CXCL13 or CXCL12 [23].

4.3.2 *Non-haematopoietic Cells*

The mesenchymal lymphoid tissue organizers (LTos) of the embryo give rise to the stromal cells of the adult SLO, FRCs of the T-cell zone, marginal reticular cells (MRCs) of the marginal zone and FDCs of the B-cell follicle. Together, they regulate organ compartmentalization, cell mobility and distribution of cells and small molecules [90].

The presence of the fibroblastic stromal cells in TLSs is determined using specific markers, such as CD35/CXCL13 for FDCs and gp38/CCL19/21 for FRCs. FDCs are generally present in TLSs that form a distinct B-cell follicle and/or a germinal centre [9]. However, because also MRCs produce CXCL13 or BAFF, the identity of FDCs must rely on discriminative markers such as CD35 for FDCs and RANKL for MRCs. FRCs, commonly characterized by the expression of gp38 and the production of CCL21, are again generally found in TLSs with a prominent T-cell recruitment [4, 80]. Krautler et al. showed that PDGFRβ+ stromal-vascular cells

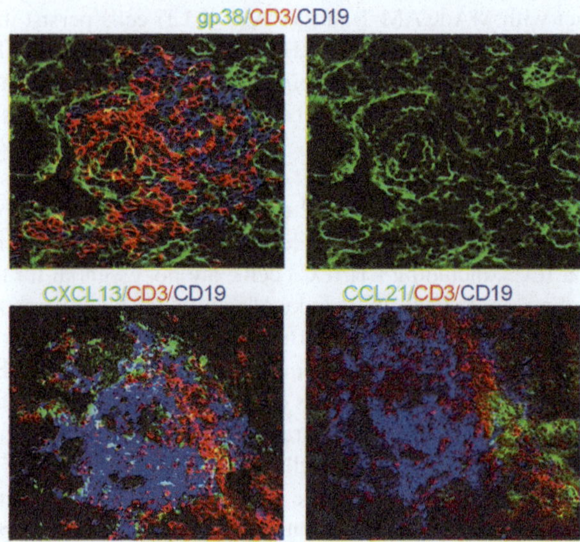

Fig. 4.2 a-d. Microphotograph illustrating TLS formation in the salivary glands of wild type mice cannulated with a replication deficient adenovirus. Staining for gp38 (green in a. and b), CD3 (red) and CD19 (blue) and CXCL13 (green in c.) and CCL21 (green in d.) illustrates the degree of lymphoid organization of the aggregates at day 15 post viral infection

from non-lymphoid organs have the capacity to differentiate into FDCs upon LTβR and TNFR triggering, suggesting that FDCs can arise at sites of stromal-vascular cells [91]. Peduto et al. also demonstrated that local resident fibroblasts give rise to immune-stromal cells in experimental models of cancer and local inflammation [92]. In mouse models of ATLSs, aorta smooth muscle cells acquired features similar to LTo expressing of VCAM-1, CXCL13 and CCL21 upon activation of the LTβR and TNFR signalling receptors. In human ATLS different types of stromal cells, including LTo-like cells, were also identified [93]. *P. aeruginosa*-induced BALT were characterized by a prominent B-cell compartment and gp38 + CXCL12+ stromal cells, while CXCL13+ FDCs did not develop [72]. Similarly in the salivary glands of mice infected with a replication-deficient adenovirus, gp38+ fibroblast differentiation is accompanied by local CXCL13 and CCL19 production ([8] and Fig. 4.2).

Analyses of chronically inflamed tissues from patients with SS or primary biliary cirrhosis have shown these tissues contain T-cell areas with reticular networks of gp38-expressing cells believed to share functional and phenotypical features of FRCs in areas rich in CCL19 and DCs [2, 94]. Similarly, CCL21+ and CXCL13+ stromal cells are present in synovial tissues of RA patients [2, 12].

SLO stromal cell differentiation is dependent on TNFSF members, and LTα/β and TNFα play an important role in immune stroma differentiation and compartmentalization in TLSs. However, LTαβ or TNFα can be dispensable for the initial events of TLS formation. In the experimental models of cancer and local inflammation by mechanical and inflammatory stimuli, the induction of LTo-like cells was shown to be independent of TNF signalling and most probably linked to the presence of polymorphonucleated cells in the first phases of the inflammatory process [92]. IL-17 directly stimulates the differentiation of CXCL13- and CCL21-expressing stromal cells [7], and lung or salivary gland fibroblasts stimulated in vitro with IL17A and IL-22 upregulated CXCL13 [6, 8]. IL-4 and IL-13 are known to activate the expression of the adhesion molecules VCAM-1 and ICAM-1 in human lung fibroblasts [95].

In a model of subcutaneous tumour apoptosis, it was found that TGF-β-induced CXCL13 expression by endogenous myofibroblasts [96]. Aorta smooth muscle cells in ATLSs also express VCAM-1, CXCL13 and CCL21 upon activation of the LTβR and TNF-RI-signalling pathways [29, 54].

Blood endothelial cells play an important role by signalling the entry of blood-derived haematopoietic cells through expression of integrins, addressins and chemokines which respond to LT$\alpha\beta$ signals [46]. Lymphatic vessels, key structures involved in leucocyte egress, expressed CCL21 and CXCL12, while the vascular endothelium stained strongly for CXCL12 in the salivary tissue of patients with SS [23]. Peri- and vascular cells express CCL21 in the rheumatoid synovium and SS [12], and CXCL13 was found on endothelial cells in salivary glands from SS patients [11]. In an inducible model of TLS formation in murine salivary glands, it has been shown that the pre-existing lymphatic vascular network undergoes expansion during TLS development, which is dependent on IL-7, LT$\alpha_1\beta_2$ and the presence of lymphocytes [97].

Although epithelial cells can transform into mesenchymal cells, there is so far no evidence that this occurs in TLS formation. In a model of skin inflammation (not characterized by full TLS maturation), stromal cells are derived from local fibroblasts but not from keratinocytes [92]. Epithelial cells appear involved in SS, where CXCL12 is expressed by the salivary duct epithelium and CXCL13 in acini and ducts [17, 23].

4.4　Treatment

Treatments aimed at depleting lymphocytes in human conditions have partially failed where established chronic TLSs were present, thus suggesting that both stromal and leucocyte components should be targeted in TLS-associated pathologies [24]. Among the therapies explored in mouse models, the administration of LTα-/β--blocking reagents has been the most used [98]. For example, in ATLS LTβR-Fc reduced TLS size concomitant with CXCL13 mRNA and B cellularity reduction and decremented HEV incidence [29]. In RIP-CXCL13 transgenic mice, LTβR-Fc led to a markedly reduction in TLS [2], but the reagent had little effect in RIP-CCL21 mice [46]. In the model of thyroiditis, LTβ-Fc inhibited HEV formation but did not disrupt lymphocyte entry [99]. Insulitis of NOD mice is effectively treated with LTβ-Fc or HVEM-Fc but not with anti-LTβ antibody [56]. LTβ-Fc has also favourable effects on SS in NOD mice [100] and in collagen-induced arthritis with a prophylactic administration [101]. LT inhibitors are currently in clinical trials for SS and RA [102, 103]. Beyond LTαβ inhibition, blocking gp38 in the model of Th17-dependent EAE appears to reduce the number of TLSs in the CNS [73]. IL-22 blocking greatly reduces TLS in a model of virus-induced SS [8], while LPS-induced iBALS was sensitive to anti-IL-17 treatment [7]. Blockade of IL-21R signalling ameliorated disease in animal models of arthritis and lupus, and an anti-human IL-21 monoclonal antibody is in clinical trials in rheumatoid arthritis and lupus [104]. Gene therapy with IL-27 in NOD mice resulted in disrupted TLS architecture, weaker antinuclear antibodies staining and improved saliva flow rates [105]

Currently, several clinical trials in autoimmune diseases target pathways described here and known to be involved in TLS formation and function, namely, IL-17, IL-21, LT, RANKL and BAFF [106].

Whether those compounds will be efficient in disaggregating TLSs in different tissues is debated, and histological results in treated humans are awaited to establish biological and clinical efficiency.

4.5 Conclusion

TLS assembly is a complex phenomenon, which can be regulated at different sites by diverse cytokines and cellular requirements. While the pathogenic versus tolerogenic role of those structures is still debated [20, 27], in chronic autoimmune disease TLSs persistence is considered a negative predictive factor for disease progression [1, 23]. Recent advances in the understanding of SLO biology and the development of novel tools to dissect leucocytes/stromal cell interaction provided critical insights in TLS assembly and regulation [7, 8]. This will translate into the development of compounds able to interfere with TLS structure and persistence in the tissue, thus decreasing local autoimmunity and the risks associated with ectopic lymphocytic expansion.

References

1. Aloisi F, Pujol-Borrell R. Lymphoid neogenesis in chronic inflammatory diseases. Nat Rev Immunol. 2006;6:205–17.
2. Link A, Hardie DL, Favre S, Britschgi MR, Adams DH, Sixt M, Cyster JG, Buckley CD, Luther SA. Association of T-zone reticular networks and conduits with ectopic lymphoid tissues in mice and humans. Am J Pathol. 2011;178:1662–75.
3. Barone F, Bombardieri M, Manzo A, Blades MC, Morgan PR, Challacombe SJ, Valesini G, Pitzalis C. Association of CXCL13 and CCL21 expression with the progressive organization of lymphoid-like structures in Sjogren's syndrome. Arthritis Rheum. 2005;52:1773–84.
4. Buckley CD, Barone F, Nayar S, Benezech C, Caamano J. Stromal cells in chronic inflammation and tertiary lymphoid organ formation. Annu Rev Immunol. 2015;33:715–45.
5. Golub R, Cumano A. Embryonic hematopoiesis. Blood Cells Mol Dis. 2013;51:226–31.

6. Khader SA, Guglani L, Rangel-Moreno J, Gopal R, Junecko BA, Fountain JJ, Martino C, Pearl JE, Tighe M, Lin YY, Slight S, Kolls JK, Reinhart TA, Randall TD, Cooper AM. IL-23 is required for long-term control of Mycobacterium tuberculosis and B cell follicle formation in the infected lung. J Immunol. 2011;187:5402–7.
7. Rangel-Moreno J, Carragher DM, De La Luz Garcia-Hernandez M, Hwang JY, Kusser K, Hartson L, Kolls JK, Khader SA, Randall TD. The development of inducible bronchus-associated lymphoid tissue depends on IL-17. Nat Immunol. 2011;12:639–46.
8. Barone F, Nayar S, Campos J, Cloake T, Withers DR, Toellner KM, Zhang Y, Fouser L, Fisher B, Bowman S, Rangel-Moreno J, Garcia-Hernandez Mde L, Randall TD, Lucchesi D, Bombardieri M, Pitzalis C, Luther SA, Buckley CD. IL-22 regulates lymphoid chemokine production and assembly of tertiary lymphoid organs. Proc Natl Acad Sci U S A. 2015;112:11024–9.
9. Bombardieri M, Barone F, Humby F, Kelly S, Mcgurk M, Morgan P, Challacombe S, De Vita S, Valesini G, Spencer J, Pitzalis C. Activation-induced cytidine deaminase expression in follicular dendritic cell networks and interfollicular large B cells supports functionality of ectopic lymphoid neogenesis in autoimmune sialoadenitis and MALT lymphoma in Sjogren's syndrome. J Immunol. 2007;179:4929–38.
10. Croia C, Astorri E, Murray-Brown W, Willis A, Brokstad KA, Sutcliffe N, Piper K, Jonsson R, Tappuni AR, Pitzalis C, Bombardieri M. Implication of Epstein-Barr virus infection in disease-specific autoreactive B cell activation in ectopic lymphoid structures of Sjogren's syndrome. Arthritis Rheumatol. 2014;66:2545–57.
11. Amft N, Curnow SJ, Scheel-Toellner D, Devadas A, Oates J, Crocker J, Hamburger J, Ainsworth J, Mathews J, Salmon M, Bowman SJ, Buckley CD. Ectopic expression of the B cell-attracting chemokine BCA-1 (CXCL13) on endothelial cells and within lymphoid follicles contributes to the establishment of germinal center-like structures in Sjogren's syndrome. Arthritis Rheum. 2001;44:2633–41.
12. Manzo A, Bugatti S, Caporali R, Prevo R, Jackson DG, Uguccioni M, Buckley CD, Montecucco C, Pitzalis C. CCL21 expression pattern of human secondary lymphoid organ stroma is conserved in inflammatory lesions with lymphoid neogenesis. Am J Pathol. 2007;171:1549–62.
13. Maglione PJ, Xu J, Chan J. B cells moderate inflammatory progression and enhance bacterial containment upon pulmonary challenge with Mycobacterium tuberculosis. J Immunol. 2007;178:7222–34.
14. Moyron-Quiroz JE, Rangel-Moreno J, Kusser K, Hartson L, Sprague F, Goodrich S, Woodland DL, Lund FE, Randall TD. Role of inducible bronchus associated lymphoid tissue (iBALT) in respiratory immunity. Nat Med. 2004;10:927–34.
15. Neyt K, Perros F, Geurtsvankessel CH, Hammad H, Lambrecht BN. Tertiary lymphoid organs in infection and autoimmunity. Trends Immunol. 2012;33:297–305.
16. Astorri E, Bombardieri M, Gabba S, Peakman M, Pozzilli P, Pitzalis C. Evolution of ectopic lymphoid neogenesis and in situ autoantibody production in autoimmune nonobese diabetic mice: cellular and molecular characterization of tertiary lymphoid structures in pancreatic islets. J Immunol. 2010;185:3359–68.
17. Salomonsson S, Larsson P, Tengner P, Mellquist E, Hjelmstrom P, Wahren-Herlenius M. Expression of the B cell-attracting chemokine CXCL13 in the target organ and autoantibody production in ectopic lymphoid tissue in the chronic inflammatory disease Sjogren's syndrome. Scand J Immunol. 2002;55:336–42.
18. Risselada AP, Looije MF, Kruize AA, Bijlsma JW, Van Roon JA. The role of ectopic germinal centers in the immunopathology of primary Sjogren's syndrome: a systematic review. Semin Arthritis Rheum. 2013;42:368–76.
19. Salomonsson S, Jonsson MV, Skarstein K, Brokstad KA, Hjelmstrom P, Wahren-Herlenius M, Jonsson R. Cellular basis of ectopic germinal center formation and autoantibody production in the target organ of patients with Sjogren's syndrome. Arthritis Rheum. 2003;48:3187–201.
20. Theander E, Vasaitis L, Baecklund E, Nordmark G, Warfvinge G, Liedholm R, Brokstad K, Jonsson R, Jonsson MV. Lymphoid organisation in labial salivary gland biopsies is a possible

predictor for the development of malignant lymphoma in primary Sjogren's syndrome. Ann Rheum Dis. 2011;70:1363–8.

21. Bugatti S, Caporali R, Manzo A, Vitolo B, Pitzalis C, Montecucco C. Involvement of subchondral bone marrow in rheumatoid arthritis: lymphoid neogenesis and in situ relationship to subchondral bone marrow osteoclast recruitment. Arthritis Rheum. 2005;52:3448–59.

22. Bugatti S, Manzo A, Benaglio F, Klersy C, Vitolo B, Todoerti M, Sakellariou G, Montecucco C, Caporali R. Serum levels of CXCL13 are associated with ultrasonographic synovitis and predict power Doppler persistence in early rheumatoid arthritis treated with non-biological disease-modifying anti-rheumatic drugs. Arthritis Res Ther. 2012;14:R34.

23. Barone F, Bombardieri M, Rosado MM, Morgan PR, Challacombe SJ, De Vita S, Carsetti R, Spencer J, Valesini G, Pitzalis C. CXCL13, CCL21, and CXCL12 expression in salivary glands of patients with Sjogren's syndrome and MALT lymphoma: association with reactive and malignant areas of lymphoid organization. J Immunol. 2008;180:5130–40.

24. Barone F, Nayar S, Buckley CD. The role of non-hematopoietic stromal cells in the persistence of inflammation. Front Immunol. 2012;3:416.

25. Hallas C, Greiner A, Peters K, Muller-Hermelink HK. Immunoglobulin VH genes of high-grade mucosa-associated lymphoid tissue lymphomas show a high load of somatic mutations and evidence of antigen-dependent affinity maturation. Lab Investig. 1998;78:277–87.

26. Qin Y, Greiner A, Hallas C, Haedicke W, Muller-Hermelink HK. Intraclonal offspring expansion of gastric low-grade MALT-type lymphoma: evidence for the role of antigen-driven high-affinity mutation in lymphomagenesis. Lab Investig. 1997;76:477–85.

27. Hu D, Mohanta SK, Yin C, Peng L, Ma Z, Srikakulapu P, Grassia G, Macritchie N, Dever G, Gordon P, Burton FL, Ialenti A, Sabir SR, Mcinnes IB, Brewer JM, Garside P, Weber C, Lehmann T, Teupser D, Habenicht L, Beer M, Grabner R, Maffia P, Weih F, Habenicht AJ. Artery tertiary lymphoid organs control aorta immunity and protect against atherosclerosis via vascular smooth muscle cell lymphotoxin beta receptors. Immunity. 2015;42:1100–15.

28. Cipponi A, Mercier M, Seremet T, Baurain JF, Theate I, Van Den Oord J, Stas M, Boon T, Coulie PG, Van Baren N. Neogenesis of lymphoid structures and antibody responses occur in human melanoma metastases. Cancer Res. 2012;72:3997–4007.

29. Grabner R, Lotzer K, Dopping S, Hildner M, Radke D, Beer M, Spanbroek R, Lippert B, Reardon CA, Getz GS, Fu YX, Hehlgans T, Mebius RE, Van Der Wall M, Kruspe D, Englert C, Lovas A, Hu D, Randolph GJ, Weih F, Habenicht AJ. Lymphotoxin beta receptor signaling promotes tertiary lymphoid organogenesis in the aorta adventitia of aged ApoE−/− mice. J Exp Med. 2009;206:233–48.

30. Magliozzi R, Columba-Cabezas S, Serafini B, Aloisi F. Intracerebral expression of CXCL13 and BAFF is accompanied by formation of lymphoid follicle-like structures in the meninges of mice with relapsing experimental autoimmune encephalomyelitis. J Neuroimmunol. 2004;148:11–23.

31. Katakai T, Hara T, Sugai M, Gonda H, Shimizu A. Th1-biased tertiary lymphoid tissue supported by CXC chemokine ligand 13-producing stromal network in chronic lesions of autoimmune gastritis. J Immunol. 2003;171:4359–68.

32. Shomer NH, Fox JG, Juedes AE, Ruddle NH. Helicobacter-induced chronic active lymphoid aggregates have characteristics of tertiary lymphoid tissue. Infect Immun. 2003;71:3572–7.

33. Manzo A, Paoletti S, Carulli M, Blades MC, Barone F, Yanni G, Fitzgerald O, Bresnihan B, Caporali R, Montecucco C, Uguccioni M, Pitzalis C. Systematic microanatomical analysis of CXCL13 and CCL21 in situ production and progressive lymphoid organization in rheumatoid synovitis. Eur J Immunol. 2005;35:1347–59.

34. Flavell RA, Kratz A, Ruddle NH. The contribution of insulitis to diabetes development in tumor necrosis factor transgenic mice. Curr Top Microbiol Immunol. 1996;206:33–50.

35. Luther SA, Lopez T, Bai W, Hanahan D, Cyster JG. BLC expression in pancreatic islets causes B cell recruitment and lymphotoxin-dependent lymphoid neogenesis. Immunity. 2000;12:471–81.

36. Martin AP, Coronel EC, Sano G, Chen SC, Vassileva G, Canasto-Chibuque C, Sedgwick JD, Frenette PS, Lipp M, Furtado GC, Lira SA. A novel model for lymphocytic infiltration of the

thyroid gland generated by transgenic expression of the CC chemokine CCL21. J Immunol. 2004;173:4791–8.

37. Bombardieri M, Barone F, Lucchesi D, Nayar S, Van Den Berg WB, Proctor G, Buckley CD, Pitzalis C. Inducible tertiary lymphoid structures, autoimmunity, and exocrine dysfunction in a novel model of salivary gland inflammation in C57BL/6 mice. J Immunol. 2012;189:3767–76.

38. Luther SA, Ansel KM, Cyster JG. Overlapping roles of CXCL13, interleukin 7 receptor alpha, and CCR7 ligands in lymph node development. J Exp Med. 2003;197:1191–8.

39. Drayton DL, Liao S, Mounzer RH, Ruddle NH. Lymphoid organ development: from ontogeny to neogenesis. Nat Immunol. 2006;7:344–53.

40. Amft N, Bowman SJ. Chemokines and cell trafficking in Sjogren's syndrome. Scand J Immunol. 2001;54:62–9.

41. Aust G, Sittig D, Becherer L, Anderegg U, Schutz A, Lamesch P, Schmucking E. The role of CXCR5 and its ligand CXCL13 in the compartmentalization of lymphocytes in thyroids affected by autoimmune thyroid diseases. Eur J Endocrinol. 2004;150:225–34.

42. Carragher DM, Rangel-Moreno J, Randall TD. Ectopic lymphoid tissues and local immunity. Semin Immunol. 2008;20:26–42.

43. Weyand CM, Goronzy JJ. Ectopic germinal center formation in rheumatoid synovitis. Ann N Y Acad Sci. 2003;987:140–9.

44. Greenbaum A, Hsu YM, Day RB, Schuettpelz LG, Christopher MJ, Borgerding JN, Nagasawa T, Link DC. CXCL12 in early mesenchymal progenitors is required for haematopoietic stem-cell maintenance. Nature. 2013;495:227–30.

45. Okada T, Ngo VN, Ekland EH, Forster R, Lipp M, Littman DR, Cyster JG. Chemokine requirements for B cell entry to lymph nodes and Peyer's patches. J Exp Med. 2002;196:65–75.

46. Luther SA, Bidgol A, Hargreaves DC, Schmidt A, Xu Y, Paniyadi J, Matlsubian M, Cyster JG. Differing activities of homeostatic chemokines CCL19, CCL21, and CXCL12 in lymphocyte and dendritic cell recruitment and lymphoid neogenesis. J Immunol. 2002;169:424–33.

47. Chen SC, Vassileva G, Kinsley D, Holzmann S, Manfra D, Wiekowski MT, Romani N, Lira SA. Ectopic expression of the murine chemokines CCL21a and CCL21b induces the formation of lymph node-like structures in pancreas, but not skin, of transgenic mice. J Immunol. 2002;168:1001–8.

48. Marinkovic T, Garin A, Yokota Y, Fu YX, Ruddle NH, Furtado GC, Lira SA. Interaction of mature CD3+CD4+ T cells with dendritic cells triggers the development of tertiary lymphoid structures in the thyroid. J Clin Invest. 2006;116:2622–32.

49. Drayton DL, Ying X, Lee J, Lesslauer W, Ruddle NH. Ectopic LT directs lymphoid organ neogenesis with concomitant expression of peripheral node addressin and a HEV-restricted sulfotransferase. J Exp Med. 2003;197:1153–63.

50. Kratz A, Campos-Neto A, Hanson MS, Ruddle NH. Chronic inflammation caused by lymphotoxin is lymphoid neogenesis. J Exp Med. 1996;183:1461–72.

51. Matsumoto M, Lo SF, Carruthers CJ, Min J, Mariathasan S, Huang G, Plas DR, Martin SM, Geha RS, Nahm MH, Chaplin DD. Affinity maturation without germinal centres in lymphotoxin-alpha-deficient mice. Nature. 1996;382:462–6.

52. Sacca R, Kratz A, Campos-Neto A, Hanson MS, Ruddle NH. Lymphotoxin: from chronic inflammation to lymphoid organs. J Inflamm. 1995;47:81–4.

53. Schrama D, Thor Straten P, Fischer WH, McLellan AD, Brocker EB, Reisfeld RA, Becker JC. Targeting of lymphotoxin-alpha to the tumor elicits an efficient immune response associated with induction of peripheral lymphoid-like tissue. Immunity. 2001;14:111–21.

54. Lotzer K, Dopping S, Connert S, Grabner R, Spanbroek R, Lemser B, Beer M, Hildner M, Hehlgans T, Van Der Wall M, Mebius RE, Lovas A, Randolph GJ, Weih F, Habenicht AJ. Mouse aorta smooth muscle cells differentiate into lymphoid tissue organizer-like cells on combined tumor necrosis factor receptor-1/lymphotoxin beta-receptor NF-kappaB signaling. Arterioscler Thromb Vasc Biol. 2010;30:395–402.

55. Gommerman JL, Browning JL. Lymphotoxin/light, lymphoid microenvironments and autoimmune disease. Nat Rev Immunol. 2003;3:642–55.

56. Lee Y, Chin RK, Christiansen P, Sun Y, Tumanov AV, Wang J, Chervonsky AV, Fu YX. Recruitment and activation of naive T cells in the islets by lymphotoxin beta receptor-dependent tertiary lymphoid structure. Immunity. 2006;25:499–509.
57. Fukuyama S, Hiroi T, Yokota Y, Rennert PD, Yanagita M, Kinoshita N, Terawaki S, Shikina T, Yamamoto M, Kurono Y, Kiyono H. Initiation of NALT organogenesis is independent of the IL-7R, LTbetaR, and NIK signaling pathways but requires the Id2 gene and CD3(−)CD4(+) CD45(+) cells. Immunity. 2002;17:31–40.
58. Harmsen A, Kusser K, Hartson L, Tighe M, Sunshine MJ, Sedgwick JD, Choi Y, Littman DR, Randall TD. Cutting edge: organogenesis of nasal-associated lymphoid tissue (NALT) occurs independently of lymphotoxin-alpha (LT alpha) and retinoic acid receptor-related orphan receptor-gamma, but the organization of NALT is LT alpha dependent. J Immunol. 2002;168:986–90.
59. Nagatake T, Fukuyama S, Kim DY, Goda K, Igarashi O, Sato S, Nochi T, Sagara H, Yokota Y, Jetten AM, Kaisho T, Akira S, Mimuro H, Sasakawa C, Fukui Y, Fujihashi K, Akiyama T, Inoue J, Penninger JM, Kunisawa J, Kiyono H. Id2-, RORgammat-, and LTbetaR-independent initiation of lymphoid organogenesis in ocular immunity. J Exp Med. 2009;206:2351–64.
60. Furtado GC, Marinkovic T, Martin AP, Garin A, Hoch B, Hubner W, Chen BK, Genden E, Skobe M, Lira SA. Lymphotoxin beta receptor signaling is required for inflammatory lymphangiogenesis in the thyroid. Proc Natl Acad Sci U S A. 2007;104:5026–31.
61. Kollias G. TNF pathophysiology in murine models of chronic inflammation and autoimmunity. Semin Arthritis Rheum. 2005;34:3–6.
62. Ernandez T, Mayadas TN. Immunoregulatory role of TNFalpha in inflammatory kidney diseases. Kidney Int. 2009;76:262–76.
63. Jacob CO, Aiso S, Michie SA, McDevitt HO, Acha-Orbea H. Prevention of diabetes in non-obese diabetic mice by tumor necrosis factor (TNF): similarities between TNF-alpha and interleukin 1. Proc Natl Acad Sci U S A. 1990;87:968–72.
64. Yu P, Lee Y, Liu W, Chin RK, Wang J, Wang Y, Schietinger A, Philip M, Schreiber H, Fu YX. Priming of naive T cells inside tumors leads to eradication of established tumors. Nat Immunol. 2004;5:141–9.
65. Bugatti S, Manzo A, Bombardieri M, Vitolo B, Humby F, Kelly S, Montecucco C, Pitzalis C. Synovial tissue heterogeneity and peripheral blood biomarkers. Curr Rheumatol Rep. 2011;13:440–8.
66. Humby F, Bombardieri M, Manzo A, Kelly S, Blades MC, Kirkham B, Spencer J, Pitzalis C. Ectopic lymphoid structures support ongoing production of class-switched autoantibodies in rheumatoid synovium. PLoS Med. 2009;6:e1.
67. Takemura S, Braun A, Crowson C, Kurtin PJ, Cofield RH, O'fallon WM, Goronzy JJ, Weyand CM. Lymphoid neogenesis in rheumatoid synovitis. J Immunol. 2001;167:1072–80.
68. Goya S, Matsuoka H, Mori M, Morishita H, Kida H, Kobashi Y, Kato T, Taguchi Y, Osaki T, Tachibana I, Nishimoto N, Yoshizaki K, Kawase I, Hayashi S. Sustained interleukin-6 signalling leads to the development of lymphoid organ-like structures in the lung. J Pathol. 2003;200:82–7.
69. Lee JJ, Mcgarry MP, Farmer SC, Denzler KL, Larson KA, Carrigan PE, Brenneise IE, Horton MA, Haczku A, Gelfand EW, Leikauf GD, Lee NA. Interleukin-5 expression in the lung epithelium of transgenic mice leads to pulmonary changes pathognomonic of asthma. J Exp Med. 1997;185:2143–56.
70. Jones GW, Bombardieri M, Greenhill CJ, McLeod L, Nerviani A, Rocher-Ros V, Cardus A, Williams AS, Pitzalis C, Jenkins BJ, Jones SA. Interleukin-27 inhibits ectopic lymphoid-like structure development in early inflammatory arthritis. J Exp Med. 2015;212:1793–802.
71. Rangel-Moreno J, Hartson L, Navarro C, Gaxiola M, Selman M, Randall TD. Inducible bronchus-associated lymphoid tissue (iBALT) in patients with pulmonary complications of rheumatoid arthritis. J Clin Invest. 2006;116:3183–94.
72. Fleige H, Ravens S, Moschovakis GL, Bolter J, Willenzon S, Sutter G, Haussler S, Kalinke U, Prinz I, Forster R. IL-17-induced CXCL12 recruits B cells and induces follicle formation in BALT in the absence of differentiated FDCs. J Exp Med. 2014;211:643–51.

73. Peters A, Pitcher LA, Sullivan JM, Mitsdoerffer M, Acton SE, Franz B, Wucherpfennig K, Turley S, Carroll MC, Sobel RA, Bettelli E, Kuchroo VK. Th17 cells induce ectopic lymphoid follicles in central nervous system tissue inflammation. Immunity. 2011;35:986–96.
74. Ciccia F, Guggino G, Rizzo A, Ferrante A, Raimondo S, Giardina A, Dieli F, Campisi G, Alessandro R, Triolo G. Potential involvement of IL-22 and IL-22-producing cells in the inflamed salivary glands of patients with Sjogren's syndrome. Ann Rheum Dis. 2012;71:295–301.
75. Lavoie TN, Stewart CM, Berg KM, Li Y, Nguyen CQ. Expression of interleukin-22 in Sjogren's syndrome: significant correlation with disease parameters. Scand J Immunol. 2011;74:377–82.
76. Meier D, Bornmann C, Chappaz S, Schmutz S, Otten LA, Ceredig R, Acha-Orbea H, Finke D. Ectopic lymphoid-organ development occurs through interleukin 7-mediated enhanced survival of lymphoid-tissue-inducer cells. Immunity. 2007;26:643–54.
77. Finke D, Acha-Orbea H, Mattis A, Lipp M, Kraehenbuhl J. CD4+CD3- cells induce Peyer's patch development: role of alpha4beta1 integrin activation by CXCR5. Immunity. 2002;17:363–73.
78. Honda K, Nakano H, Yoshida H, Nishikawa S, Rennert P, Ikuta K, Tamechika M, Yamaguchi K, Fukumoto T, Chiba T, Nishikawa SI. Molecular basis for hematopoietic/mesenchymal interaction during initiation of Peyer's patch organogenesis. J Exp Med. 2001;193:621–30.
79. Mebius RE. Organogenesis of lymphoid tissues. Nat Rev Immunol. 2003;3:292–303.
80. Pitzalis C, Jones GW, Bombardieri M, Jones SA. Ectopic lymphoid-like structures in infection, cancer and autoimmunity. Nat Rev Immunol. 2014;14:447–62.
81. Rangel-Moreno J, Moyron-Quiroz JE, Hartson L, Kusser K, Randall TD. Pulmonary expression of CXC chemokine ligand 13, CC chemokine ligand 19, and CC chemokine ligand 21 is essential for local immunity to influenza. Proc Natl Acad Sci U S A. 2007;104:10577–82.
82. Sato M, Hirayama S, Matsuda Y, Wagnetz D, Hwang DM, Guan Z, Liu M, Keshavjee S. Stromal activation and formation of lymphoid-like stroma in chronic lung allograft dysfunction. Transplantation. 2011;91:1398–405.
83. Yoshida H, Naito A, Inoue J, Satoh M, Santee-Cooper SM, Ware CF, Togawa A, Nishikawa S. Different cytokines induce surface lymphotoxin-alphabeta on IL-7 receptor-alpha cells that differentially engender lymph nodes and Peyer's patches. Immunity. 2002;17:823–33.
84. Furtado GC, Pacer ME, Bongers G, Benezech C, He Z, Chen L, Berin MC, Kollias G, Caamano JH, Lira SA. TNFalpha-dependent development of lymphoid tissue in the absence of RORgammat(+) lymphoid tissue inducer cells. Mucosal Immunol. 2014;7:602–14.
85. Gunn MD, Ngo VN, Ansel KM, Ekland EH, Cyster JG, Williams LT. A B-cell-homing chemokine made in lymphoid follicles activates Burkitt's lymphoma receptor-1. Nature. 1998;391:799–803.
86. Browning JL, Allaire N, Ngam-Ek A, Notidis E, Hunt J, Perrin S, Fava RA. Lymphotoxin-beta receptor signaling is required for the homeostatic control of HEV differentiation and function. Immunity. 2005;23:539–50.
87. Moussion C, Girard JP. Dendritic cells control lymphocyte entry to lymph nodes through high endothelial venules. Nature. 2011;479:542–6.
88. Suematsu S, Watanabe T. Generation of a synthetic lymphoid tissue-like organoid in mice. Nat Biotechnol. 2004;22:1539–45.
89. Geurtsvankessel CH, Willart MA, Bergen IM, Van Rijt LS, Muskens F, Elewaut D, Osterhaus AD, Hendriks R, Rimmelzwaan GF, Lambrecht BN. Dendritic cells are crucial for maintenance of tertiary lymphoid structures in the lung of influenza virus-infected mice. J Exp Med. 2009;206:2339–49.
90. Roozendaal R, Mebius RE. Stromal cell-immune cell interactions. Annu Rev Immunol. 2011;29:23–43.
91. Krautler NJ, Kana V, Kranich J, Tian Y, Perera D, Lemm D, Schwarz P, Armulik A, Browning JL, Tallquist M, Buch T, Oliveira-Martins JB, Zhu C, Hermann M, Wagner U, Brink R, Heikenwalder M, Aguzzi A. Follicular dendritic cells emerge from ubiquitous perivascular precursors. Cell. 2012;150:194–206.

92. Peduto L, Dulauroy S, Lochner M, Spath GF, Morales MA, Cumano A, Eberl G. Inflammation recapitulates the ontogeny of lymphoid stromal cells. J Immunol. 2009;182:5789–99.
93. Dutertre CA, Clement M, Morvan M, Schakel K, Castier Y, Alsac JM, Michel JB, Nicoletti A. Deciphering the stromal and hematopoietic cell network of the adventitia from non-aneurysmal and aneurysmal human aorta. PLoS One. 2014;9:e89983.
94. Drumea-Mirancea M, Wessels JT, Muller CA, Essl M, Eble JA, Tolosa E, Koch M, Reinhardt DP, Sixt M, Sorokin L, Stierhof YD, Schwarz H, Klein G. Characterization of a conduit system containing laminin-5 in the human thymus: a potential transport system for small molecules. J Cell Sci. 2006;119:1396–405.
95. Doucet C, Brouty-Boye D, Pottin-Clemenceau C, Canonica GW, Jasmin C, Azzarone B. Interleukin (IL) 4 and IL-13 act on human lung fibroblasts. Implication in asthma. J Clin Invest. 1998;101:2129–39.
96. Ammirante M, Shalapour S, Kang Y, Jamieson CA, Karin M. Tissue injury and hypoxia promote malignant progression of prostate cancer by inducing CXCL13 expression in tumor myofibroblasts. Proc Natl Acad Sci U S A. 2014;111:14776–81.
97. Nayar S, Campos J, Chung MM, Navarro-Nunez L, Chachlani M, Steinthal N, Gardner DH, Rankin P, Cloake T, Caamano JH, Mcgettrick HM, Watson SP, Luther S, Buckley CD, Barone F. Bimodal expansion of the lymphatic vessels is regulated by the sequential expression of IL-7 and lymphotoxin alpha1beta2 in newly formed tertiary lymphoid structures. J Immunol. 2016;197:1957–67.
98. Browning JL. Inhibition of the lymphotoxin pathway as a therapy for autoimmune disease. Immunol Rev. 2008;223:202–20.
99. Marinkovic T. Interaction of mature CD3+CD4+ T cells with dendritic cells triggers the development of tertiary lymphoid structures in the thyroid. J Clin Investig. 2006;116:2622–32.
100. Fava RA, Kennedy SM, Wood SG, Bolstad AI, Bienkowska J, Papandile A, Kelly JA, Mavragani CP, Gatumu M, Skarstein K, Browning JL. Lymphotoxin-beta receptor blockade reduces CXCL13 in lacrimal glands and improves corneal integrity in the NOD model of Sjogren's syndrome. Arthritis Res Ther. 2011;13:R182.
101. Fava RA, Notidis E, Hunt J, Szanya V, Ratcliffe N, Ngam-Ek A, De Fougerolles AR, Sprague A, Browning JL. A role for the lymphotoxin/LIGHT axis in the pathogenesis of murine collagen-induced arthritis. J Immunol. 2003;171:115–26.
102. Gatumu MK, Skarstein K, Papandile A, Browning JL, Fava RA, Bolstad A. Blockade of lymphotoxin-beta receptor signaling reduces aspects of Sjögren syndrome in salivary glands of non-obese diabetic mice. Arthritis Res Ther. 2009;11:R24.
103. Ruddle NH. Lymphotoxin and TNF: how it all began-a tribute to the travelers. Cytokine Growth Factor Rev. 2014;25:83–9.
104. Herber D, Brown TP, Liang S, Young DA, Collins M, Dunussi-Joannopoulos K. IL-21 has a pathogenic role in a lupus-prone mouse model and its blockade with IL-21R.Fc reduces disease progression. J Immunol. 2007;178:3822–30.
105. Lee BH, Carcamo WC, Chiorini JA, Peck AB, Nguyen CQ. Gene therapy using IL-27 ameliorates Sjogren's syndrome-like autoimmune exocrinopathy. Arthritis Res Ther. 2012;14:R172.
106. Bombardieri M, Lewis M, Pitzalis C. Ectopic lymphoid neogenesis in rheumatic autoimmune diseases. Nat Rev Rheumatol. 2017;13:141–54.

Chapter 5
Mesenchymal Stem Cells as Endogenous Regulators of Inflammation

Hafsa Munir, Lewis S. C. Ward, and Helen M. McGettrick

Abstract This chapter discusses the regulatory role of endogenous mesenchymal stem cells (MSC) during an inflammatory response. MSC are a heterogeneous population of multipotent cells that normally contribute towards tissue maintenance and repair but have garnered significant scientific interest for their potent immunomodulatory potential. It is through these physicochemical interactions that MSC are able to exert an anti-inflammatory response on neighbouring stromal and haematopoietic cells. However, the impact of the chronic inflammatory environment on MSC function remains to be determined. Understanding the relationship of MSC between resolution of inflammation and autoimmunity will both offer new insights in the use of MSC as a therapeutic, and also their involvement in the pathogenesis of inflammatory disorders.

Keywords Mesenchymal stem cells · Endothelial cells · Neutrophils · Lymphocytes

5.1 Introduction

Mesenchymal stem cells (MSC) are non-haematopoietic, multipotent tissue-resident precursor cells with immunomodulatory capabilities [1]. They exist in small numbers in a variety of tissues including the bone marrow (BM), Wharton's jelly (WJ), adipose tissue (AD), dental pulp, brain, and spleen [2]. Even within different tissues, MSC are thought to exhibit heterogeneous phenotypes based on cellular size,

Hafsa Munir and Lewis S. C. Ward have contributed equally to this work

H. Munir
MRC Cancer Unit/Hutchison, University of Cambridge, Cambridge, UK

L. S. C. Ward
Discovery Sciences, AstraZeneca, Cambridge, UK

H. M. McGettrick (✉)
Rheumatology Research Group, Institute of Inflammation and Ageing,
University of Birmingham, Birmingham, UK

© Springer International Publishing AG, part of Springer Nature 2018
B. M.J Owens, M. A. Lakins (eds.), *Stromal Immunology*, Advances in Experimental Medicine and Biology 1060, https://doi.org/10.1007/978-3-319-78127-3_5

surface marker expression, differentiation capacity, and function [3–6]. Thus, not all MSC are the same. Indeed, growing evidence suggests that the MSC niche is unique in distinct tissues and that variation in tissue microenvironments may lead to tissue-specific differences in MSC functions [7–10]. As well as their reparative roles, MSC possess immunomodulatory capabilities and therefore have the potential to regulate inflammation and its resolution. MSC-mediated immunomodulation occurs through two mechanisms: release of soluble factors and cell-cell contact-dependent interactions (Table 5.1). Here, we review the origins of tissue-resident MSC, their interaction with the tissue microenvironment, and how this may influence inflammatory responses. A brief synopsis on MSC as a therapeutic strategy for the treatment of graft-versus-host disease is also discussed.

Table 5.1 Immunomodulatory effects of MSC on haematopoietic and stromal cells

Affected cell	Effect	Mediator(s)	Species	Passage	References
Stem cells					
HSC	↓ BM egress	CXCL12	Mouse	–	[11–13]
	↑ Proliferation and maintain HSC in an undifferentiated state	β-catenin	Mouse	–	[14, 15]
Leukocytes					
Neutrophils	↑ Phagocytosis	Soluble factors	Human	3–5	[16]
	↓ Respiratory burst and apoptosis	Soluble factors	Human	3–5	[16, 17]
NK cells	↓ IFN‖ secretion and cytotoxicity	PGE$_2$, HLA-G5	Human	1–6	[18–20]
Monocytes	↓ IL-12 secretion	PGE$_2$	Human	≥2–4	[21, 22]
	↑ BM egress	CCL2	Mouse	–	[23]
	↓ Differentiation into DC	IL-6, M-CSF, PGE$_2$	Human	≤15	[21, 22]
	↑ Polarisation to M2 macrophage	IDO, PGE$_2$	Human/ mouse	3–7	[24–26]
T-cells	↓ Proliferation	TGFβ, HGF, PD-1-PD-L1/2, NO, PGE$_2$	Human/ mouse	1–6	[21, 27–37]
	↓ IFNγ secretion	Cell contact, IL-10	Human	≤6	[38, 39]
	↑ Expansion of T$_{reg}$	HLA-G	Human	1	[30]
B-cells	↓ Antibody production	Soluble factors	Human	–	[40]
	↓CXCR4, CXCR5, CCR7 expression inhibiting trafficking	Soluble factors	Human	–	[40]
	↓ Proliferation	Cell contact	Human/ mouse	–	[32, 40]
DC	↓ TNFα secretion	IL-10	Human	≤6	[38]
	↓ Antigen-presenting functions	–	Human/ mouse	–	[22, 39, 41]
	↓ CCR7 expression ↓ trafficking	Soluble factors	Human	–	[42]

(continued)

Table 5.1   (continued)

Affected cell	Effect	Mediator(s)	Species	Passage	References
Stromal cells					
Endothelial cells	↑ Proliferation and migration	CCL2,CXCL12VEGF, PDGF	Human/ rodent	3–5	[43–45]
	↑ Angiogenesis	ROS	Rat	–	[46]
	↓ Vascular permeability	S-1-P	Human	3–7	[47–50]
	↓ Leukocyte recruitment[a]	IL-6, TGFβ	Human	3	[51–53]

All behaviours were analysed with BMMSC

[a]Also analysed for WJ MSC

IDO indoleamine 2,3-dioxygenase, *PD1* programmed cell death 1, *PGE2* prostaglandin E2, *ROS* reactive oxygen species, *S-1-P* sphingosine-1-phosphate

5.2   Origin of MSC

Our best definition of an MSC is defined in the International Society for Cell Therapy 2006 guidelines (Fig. 5.1) [54]. Additional surface proteins (e.g. CD146 and CD271) are thought to identify highly potent (suppressive) MSC subpopulations as assessed by T-cell proliferation assays [55]. Despite this, no specific MSC marker – based on either surface expression or function – has been identified. Moreover, "MSC" markers are also found on non-MSC stromal populations (e.g. fibroblasts) indicating that this criterion is too generic for defining a specific population in tissue. Also of concern is that the morphology, differentiation capacity, and expression of "MSC" markers are modified to varying degrees by in vitro culture conditions [56]. Identification of a unique, functionally relevant marker is urgently required to truly elucidate the endogenous role of tissue-resident MSC in modulating inflammation and the effects of MSC therapy in vivo. Understanding the origin of MSC may identify early lineage-specific markers that are exclusively expressed on MSC and can be used to distinguish these cells from other stromal cells.

Little is known about the developmental origin of MSC, with recent evidence suggesting at least two distinct lineages: neural crest and mesoderm. MSC can differentiate into cells of the neural lineages, and subsets of murine BM-derived MSC have been reported to express neural crest stem cell-specific genes [57], leading several groups to postulate this as their origin [57, 58]. Additionally, murine neural crest-derived cells can migrate through the bloodstream to populate numerous tissues, including the bone marrow, where they exhibit a differentiation capacity indicative of stem cells [58]. In contrast, lineage tracing studies showed that cells from the primary vascular plexus give rise to perivascular cells that exhibit MSC-like properties [59–61]. Whilst the origin of MSC is still being debated, it is clear that the cells described in these studies exhibit the same phenotypic features of MSC in vitro. Identifying the origin of MSC and their organ distribution (i.e. differences between MSC populations) may explain functional variations observed in MSC isolated from different anatomical sites.

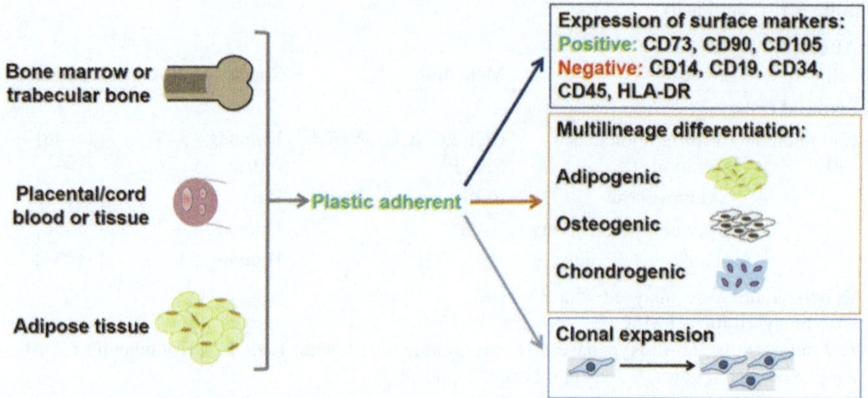

Fig. 5.1 Definition for mesenchymal stem cells. MSC can be isolated from a variety of sources (bone marrow, placenta/umbilical cord, and adipose tissue) primarily based on plastic adherence. Due to the heterogeneity of these cells, further characterisation is required. The International Society for Cell Therapy described the minimum criteria necessary to define MSC [54]. The cells must express the stromal markers, CD73, CD90, and CD105, and lack expression of haematopoietic and endothelial markers, CD14, CD19, CD34, CD45, and HLA-DR. They must also be able to differentiate into other mesodermal lineages (adipogenic, osteogenic, and chondrogenic). Lastly, MSC must be able to undergo clonal expansion during in vitro culture

5.3   MSC in the Bone Marrow Niche

BMMSC can contribute to the haematopoietic stem cell (HSC) niche by regulating haematopoiesis [11, 14, 15] and trafficking of BM-derived cells into the circulation [11–13]. Depletion of MSC or MSC-like progenitors caused an increase in HSC mobilisation [11] and augmented the expression of early myeloid selector genes by HSC, reducing their overall number in the bone marrow [15]. This indicates that the presence of MSC in the HSC niche is essential for inducing their proliferation and maintaining HSC in an undifferentiated state [15]. Indeed, stimulation of β-catenin in MSC has been shown to promote HSC self-renewal in vivo suggesting that this signalling pathway is involved [14]. MSC can also "hold" HSC in the perivascular niche through CXCL12-CXCR4-dependent interactions, preventing them from exiting the bone marrow into the bloodstream, akin to the mechanism reported for mature leukocytes [11, 12]. Importantly, the expression of CXCL12 by MSC can be regulated by CD169+ macrophages within the BM niche [13]. Depleting these BM macrophages reduced CXCL12 expression on MSC and in turn enhanced HSC egress [13]. Thus, MSC play an integral role in maintaining HSC within the BM niche through soluble mediators but also complex multicellular cross-talk with HSC and mature leukocytes.

Evidence suggests that MSC may also regulate the trafficking of monocytes and B cells from the bone marrow [13, 23]. During systemic infection, BMMSC up-regulated CCL2 in response to toll-like receptor (TLR) activation, promoting the egress of CCR2$^+$ monocytes into the bloodstream [23]. This mobilisation of monocytes also promotes HSC egress away from the stem cell niche [13, 23] encouraging their maturation into leukocytes. This tightly regulated process requires cross-talk between MSC, monocytes, and HSC to coordinate an appropriate immune response. BMMSC also down-regulated expression of CXCR4 by B cells, which may promote their exit from the bone marrow [40]. Whether MSC influence maturation of other leukocyte populations remains to be determined (reviewed by [62]). The main function of BM-resident MSC is to endogenously regulate the proliferation and maturation of HSC and may therefore indirectly influence leukocyte generation. Additionally, MSC may also regulate leukocyte egress in response to infection and/or inflammatory cues. This indicates a novel and potentially tissue-specific role of BM-resident MSC.

5.4 MSC Regulation of Immune Cells

5.4.1 Effects on Innate Immunity

Within the tissue, resident MSC are thought to modulate the movement, effector functions, and survival of recruited neutrophils. Several studies have reported enhanced neutrophil chemotaxis across blank filters towards conditioned media from resting MSC, lipopolysaccharide (LPS)-primed MSC, or MSC isolated from diseased tissue (e.g. gastric cancer) [16, 63, 64]. However, direct coculture of MSC with neutrophils for 1 h, in contrast, had no effect on the ability of neutrophils to migrate along a gradient of C5a, IL-8, or fMLP [17]. In conflicting studies, BMMSC have been shown to dampen the fMLP-induced respiratory burst of neutrophils [17], whilst supernatants from BMMSC enhanced oxidative release in LPS-primed neutrophils [16]. Indeed, these supernatants were also demonstrated to augment neutrophil phagocytosis [16]. Furthermore, coculture with BMMSC or WJMSC or supernatants from parotid gland MSC reduced neutrophil apoptosis in vitro at 18–24 h [16, 17, 65]. Certain contexts require cell-cell contact in conjunction with soluble mediators to elicit the effects of MSC; however the reasons for this remain unknown. One possibility is that these rely on similar mechanisms to those observed with ICAM1-mediated suppression in lymphocytes [18, 19], but further investigations are required.

MSC have also been reported to dampen innate immune responses by suppressing the effector functions of natural killer (NK) cells and skewing the differentiation of monocytes towards a more anti-inflammatory M2 phenotype [20]. Human BMMSC suppressed IFNγ secretion by IL-2 [21, 38] or IL-15 [66] activated NK cells. In the case of the latter study, this was partially mediated through prostaglandin E$_2$ [PGE$_2$] and to a lesser extent TGFβ [66]. Cytotoxic effector functions of activated NK cells are also suppressed by BMMSC in vitro [21, 66] via indolamine-2,3-dioxygenase [IDO] and PGE$_2$ acting synergistically [21]. Similarly, contact with BMMSC also promoted monocyte polarisation to IL-10 producing M2 macrophages, once again in a soluble mediator (IDO and PGE$_2$)-dependent manner [24–26]. Indeed, IL-10 produced from M2 macrophages reduced neutrophil infiltration and lethality of sepsis in vivo following infusion of BMMSC [67]. In contrast, human BMMSC can suppress allogeneic CD14$^+$ monocyte differentiation into dendritic cells in vitro (driven by GM-CSF, IL-4, and LPS) when cells were cultured in close proximity, but not direct contact, on opposite sides of a porous filter [22]. MSC appear to have the ability to "turn off" inflammatory responses promoting resolution. Indeed preconditioning U937 cells (monocytic cell line) with BMMSC for 16 h reduced their adhesion to inflamed pulmonary endothelial cells in vitro [68]. Thus, tissue-resident MSC may act as endogenous sensors of inflammation, influencing the activity of recruited leukocytes. Moreover, they may also coordinate the switch from innate to adaptive immunity during protective inflammation.

5.4.2 Effects on Adaptive Immunity

MSC modulation of T-cell behaviour has been extensively studied (reviewed by [27]). MSC from a variety of tissues promote the survival of T-cells whilst maintaining them in a quiescent state by suppressing proliferation [28–30] and the production of pro-inflammatory cytokines (e.g. IFNγ) [38]. Indeed, these represent the standard assays used to test the potency of MSC. As with other cell types, MSC mediate their effects through soluble factors (e.g. TGF-β, IDO, and PGE2) and cell contact (e.g. programmed cell death 1 [PD-1]) (reviewed by [69]). These factors can synergistically induce maximal suppression of T-cell proliferation when MSC are in direct contact with the T-cells [31]. Cell-cell contact between MSC and T-cells leads to bidirectional cross-talk affecting both cell types. For example, ICAM-1 is up-regulated by human ADMSC following interaction with T-cells and is necessary for the suppression of proliferation, where blocking ICAM-1 on ADMSC releases T-cells from IDO-induced inhibition [70]. BMMSC can also enhance the expansion of the T$_{reg}$ population in peripheral blood mononuclear cells in a HLA-G-dependent manner, which may be further enhanced by IL-10 [30]. Moreover, human ADMSC have been shown to redirect B-cell plasmablast formation into a regulatory B-cell subset (B$_{reg}$), although the mechanism remains unknown [71, 72]. Consequently, MSC could potentially amplify their effects on T-cells indirectly, by promoting the proliferation of local T$_{reg}$ and B$_{reg}$ populations.

How MSC regulate other cells of the adaptive immune system is poorly understood. Human BMMSC have been reported to preserve naive B-cells in a resting state suppressing their proliferation and antibody production [19, 40]. Similar observations have been made in mice where BMMSC inhibited the expansion of follicular and marginal zone B-cells in vitro [73]. Coculture in contact with MSC reduced the expression of chemokine receptors on B-cells (CXCR5 and CCR7) and dendritic cells (CCR7; [42]) required for trafficking through lymphoid organs [40]. Additionally MSC are capable of promoting tolerance in vitro: coculture on opposite sides of a porous filter impaired NF-κB signalling in dendritic cells resulting in reduced CD80/CD86 and HLA expression and impaired stimulation of T-cell clonal expansion [22, 39, 41, 74]. In contrast data from phase I to phase II clinical trials in patients undergoing liver transplants has observed no tolerogenic effect of BMMSC infusion [75]. In most cases MSC-derived agents are sufficient to drive their effects on adaptive immune cells. However in a few cases, direct cell contact appeared necessary to produce a maximal response possibly involving the PD-1 pathway [32, 73].

5.5 MSC Interactions with Platelets

MSC are also capable of interacting with circulating platelets. Whilst we know much less about these interactions, they are likely to be critically important in the context of MSC cell-based therapy and vascular damage where perivascular MSC become exposed to blood [59, 60]. Human MSC bind circulating platelets in a β_1-integrin-dependent manner [76], where such interactions enhanced MSC adhesion to arterial endothelium in vitro [77] and facilitated BMMSC recruitment to lung vasculature in a rat model of pulmonary arterial hypertension [78]. Similarly platelet-MSC interactions also impact the ability of the MSC therapy to bind to extracellular matrix proteins such as collagen and fibronectin [76]. Furthermore, depleting platelets have been shown to impair MSC homing, a murine model of LPS-induced dermal inflammation [79]. Collectively these studies indicate that platelet-MSC interactions may aid their "homing" to damaged sites following therapeutic administration. However, caution is required as recent evidence indicates that such interactions have the potential to induce platelet activation and cause thrombus formation. The glycoprotein podoplanin, which is expressed by human WJMSC, can bind to CLEC-2 on platelets and induce platelet activation and their subsequent aggregation [76]. When administered systemically, podoplanin-expressing WJMSC cause a significant reduction in platelet numbers in the blood, with the platelets forming higher-order aggregates of activated cells [76]. Thus, platelet-MSC interactions have the potential to be beneficial in facilitating MSC homing to inflammatory sites but also detrimental associated with increased the risk of thrombotic events. Further investigations are required to resolve the functional impact of MSC on platelets and vice versa.

5.6 MSC Regulation of Vascular Endothelial Cells and Tissue-Resident Stroma

MSC reside in the perivascular niche in close proximity with endothelial cells (EC) lining the vasculature (blood and lymphatic) and other tissue-resident (stromal) cells [59, 60]. Comparatively speaking we understand very little about the interactions of MSC with these populations and their functional consequences. Indeed the effects of MSC on the behaviour of endothelial cells have been analysed in three contexts (see below), whilst their interactions with stromal cells have solely focused on the reparative properties of both cell types.

5.6.1 Regulation of Angiogenesis

Under resting conditions, human and rodent BMMSC have been reported to release factors (e.g. VEGFα and PDGF-BB) known to enhance the proliferation and migration of endothelial cells [43–45]. The production of these agents indicates that MSC have the potential to promote angiogenesis. In a murine model of wound repair, BMMSC (injected intradermally) and BMMSC-derived conditioned media (injected subcutaneously at the site of injury) increased endothelial cell and macrophage numbers at the site of the wound [44, 80]. These studies suggest that MSC promote wound healing by inducing angiogenesis. In vitro, proliferation and migration of both human and murine endothelial cells was induced in the presence of conditioned media from BMMSC but not dermal fibroblasts [44]. For further information on the effects MSC have on in vitro tube-forming assays, see review [81]. Of note, the main stimulators of angiogenesis, like shear stress and oxygen tension, were not modelled in these studies. Furthermore, co-injection of MSC with B16 melanoma cells increased tumour size and vessel area in vivo, indicating that they are pro-angiogenic [82]. In contrast, MSC suppressed angiogenesis in a Matrigel model through production of reactive oxygen species when in direct contact with rat lung microvascular EC [46]. Whether these factors are the key drivers of MSC-induced angiogenesis has not been explored. Numerous putative angiogenic proteins have recently been identified in exosomes derived from MSC cultured under serum-starved hypoxic conditions [83]. MSC-derived factors may well communicate with endothelial cells to control angiogenesis during development and wound repair. Endogenous MSC regulation of angiogenesis in adult pathologies remains unclear.

5.6.2 Regulation of Blood Vascular Permeability

Evidence suggests that perivascular MSC can communicate with endothelial cells to regulate vascular permeability and maintain vessel integrity in resting and acute inflammatory conditions [47–50, 84]. Coculture with MSC increased the stability of junctional molecules (e.g. VE-cadherin and β-catenin) by inhibiting their turnover at the plasma membrane of endothelial cells, reducing endothelial permeability to FITC-dextran [50]. This effect was reproduced when endothelial cells were treated with conditioned media from the coculture, implicating soluble mediators as the main drivers [50]. In LPS-driven infection, infusion of BMMSC reduced pulmonary microvessel permeability and increased endothelial barrier function in vivo, reducing murine lung vascular permeability [49]. Similar observations were made using both mouse and rat models of haemorrhagic shock [47, 84]. Nevertheless, therapeutic administration of MSC may have beneficial effects for individuals with severe vascular damage.

5.6.3 Regulation of Leukocyte Recruitment

In terms of regulating inflammatory responses, perivascular MSC communicate directly with neighbouring endothelium to indirectly regulate leukocyte recruitment during inflammation [47, 51, 68]. However, very few studies have examined this, and none have questioned whether MSC from different tissues have the same capacity to regulate this process (i.e. tissue-specific effects).

Therapeutic administration of murine BMMSC increased the number of circulating neutrophils whilst simultaneously decreasing circulating monocytes in a murine model of sepsis, suggesting MSC can actively influence leukocyte recruitment [67]. Moreover, pretreating pulmonary endothelial cells with conditioned media from human endothelial-BMMSC cocultures reduced their ability to support monocytic leukaemia cell line (U937) adhesion in response to TNFα in vitro, by tightening endothelial adherens junctions (VE-cadherin and β-catenin) and reducing adhesion molecule expression, ICAM-1 and VCAM-1 [47]. Thus, MSC can reduce leukocyte adhesion when they interact directly with target cells. However, these studies analysed adhesion under static conditions, which do not mimic physiological recruitment of leukocytes from flowing blood. Moreover, they focus on soluble mediator-induced effects on naive endothelium, rather than the direct bidirectional cross-talk between MSC and endothelial cells.

To address this, we developed an in vitro multicellular flow-based adhesion assay that mimicked intravenous BMMSC and WJMSC infusion and subsequent integration into the endothelial monolayer [51, 52]. We reported that MSC communicate with neighbouring vascular endothelial cells to limit leukocyte recruitment induced by inflammatory cytokines [51, 53]. Specifically, BMMSC potently down-regulated the recruitment of both neutrophils and lymphocytes by inflamed endothelium [51, 53]. Whilst WJMSC and TBMSC elicited similar effects, these MSC populations showed greater suppressive effects compared to BMMSC, which could be attributed to tissue-specific differences [51, 53]. A two-way conversation between MSC and endothelial cells was essential for these effects, with activation of TGFβ and release of IL-6 being critical factors [51, 53]. Coculture with MSC also inhibited the secretion of chemokines (CXCL8 and CXCL10) responsible for stabilising leukocyte adhesion and driving onward migration [51].

Alternatively, MSC and endothelial cells were cocultured together on opposite sides of a porous insert. This construct more accurately models the cross-talk that occurs within the tissue but can also be used to examine the effects of site-specific infusion of MSC [52, 53]. Like the therapeutic model, we observed that BMMSC and WJMSC suppressed neutrophil recruitment. Once again, coculture conditioned media mimicked the effects of coculture, indicating a soluble mediator-dependent mechanism. Indeed, IL-6 and TGFβ were identified as the main mediators. Interestingly, production of the soluble mediator by WJMSC, but not BMMSC, was dependent upon close proximity between the MSC and EC [53]. This suggests that BMMSC can communicate with endothelial cells in a contact-independent manner [53]. We have shown that MSC communicate directly with neighbouring endothelium to modulate the inflammatory response. Whilst MSC from different anatomical sites have the same functional effects, they appear to utilise different mechanisms which may ultimately affect their regulatory capacity. These functional differences may be due to differences in developmental origin of different MSC populations, a phenomenon previously observed in different smooth muscle cell populations [85]. This has important implications for therapy, as it suggests that MSC from different sources may only suppress recruitment when administered in close proximity to the endothelium.

These observations are not restricted to tissue-resident MSC. We and others have shown that healthy stromal cells from a variety of tissues (e.g. fibroblasts, podocytes, and secretory smooth muscle cells) exhibit immunosuppressive capabilities, limiting leukocyte recruitment induced by inflammatory cytokines [[51, 86–88]; also see Chap. 3]. Moreover, stromal populations, including endothelial cells and fibroblasts, display distinct spatial identities [89] that govern their behaviour. This allows them to establish tissue-specific "address codes" that actively regulate the recruitment of leukocytes to inflamed sites (reviewed by [90]). Whether MSC exhibit such tissue-specific differences requires further investigation. Collectively these studies suggest that healthy mesenchymal tissue-resident cells use the same mechanism to act as endogenous regulators of the inflammatory infiltrate, with IL-6 and TGFβ acting as master regulators [51, 53]. Given these agents are present in endothelial-MSC conditioned media, infusion of culture supernatant or MSC-derived agents may be more efficacious than infusion of cells. Ultimately this would eliminate the need for MSC infusions where the long-term effects (safety and efficacy) of therapy are unknown.

5.6.4 Regulation of Tissue Repair: Interactions with Stromal Cells

Limited evidence suggests MSC may interact with other tissue-resident mesenchymal stromal cells to facilitate their reparative functions during tissue repair and bone remodelling [91–95]. BMMSC have been reported to migrate towards damaged bone in response to TGFβ1 released by osteoclastic bone at resorptive sites, where they differentiate into osteoblasts promoting bone remodelling [91]. Moreover, rheumatoid synovial fibroblasts secrete placental growth factor, promoting BMMSC chemotaxis [96]. In rodent models of tissue damage (surgically or chemically induced), injection of BMMSC or BMMSC conditioned media reduced tissue fibrosis in the affected organ (kidney, heart, liver, and skin; [92–95]). One interpretation is that MSC migrate into the damaged tissue to communicate with resident fibroblasts and influence their production and/or deposition of extracellular matrix components, reducing fibrosis. Indeed, Yates et al. have recently demonstrated that MSC and fibroblasts can synergistically reduce extracellular matrix production and thus scarring when transplanted into a CXCR3-deficient mouse model [97]. New lines of research are necessary to determine whether MSC manipulate stroma responses to regulate the tissue microenvironment during inflammation.

5.7 Regulation by the Physical Microenvironment

MSC respond to nanoscale features altering their growth and differentiation potentials according to the patterns of nanotopography they experience [98]. For example, soft (0.5 kPa) hydrogels promoted MSC differentiation towards neural cells, whilst stiff (40 kPa) gels drive osteogenesis in the absence of additional growth factors [99]. Moreover, MSC pluripotency can be maintained using a highly ordered distribution of nanopits on the culture surface [100]. Introducing a relatively small amount of disorder to such features was sufficient to stimulate osteogenesis [101]. Sensing topographical features smaller than adhesion molecules (~10 nm) indicates that MSC observe fine details (physical and chemical) within their environment and are able to mount potent responses in an effort to maintain tissue homeostasis. Such insights could enable the ex vivo expansion of MSC for therapeutic use on specially designed surfaces that can topographically maintain, e.g. "stemness".

5.8 MSC Response to Acute Inflammation

The inflammatory microenvironment is complex with a context-specific medley of agents that can shape the behaviour of leukocytes, endothelial cells, and stromal cells. Do tissue-resident MSC also respond to their local environment and does this impact their effector functions?

One avenue that has been explored is the effects of exogenous cytokines on the phenotype of MSC (Table 5.2) and the functional consequences of these changes (Table 5.3). Pretreating MSC (BM, WJ, AD) with IFNγ in combination with TNFα for 18 h altered their phenotype: differentially modifying TLR expression (see Table 5.2) and increasing the release of cytokines (e.g. IL-6) and chemokines (e.g. CXCL8, CCL5) when compared to untreated MSC [104]. Murine BMMSC treated with IFNγ in combination with either TNFα, IL-1α, or IL-1β for 24 h up-regulated expression of adhesion molecules (e.g. ICAM1, VCAM1) and chemokine (e.g. CXCL9) compared to untreated MSC [18, 106]. Of note, single cytokine treatments had little effect on these parameters [18, 106]. In contrast, IFN||, but not TNF||, stimulation for 72 h induced IDO expression by BMMSC and WJMSC relative to resting MSC [102]. Many of these changes mirror the response of other stromal cell types to inflammation ([114, 115]; see Chap. 3) and support cell-cell interactions necessary for migration to the damaged tissue. In certain contexts, cytokines can further enhance the immunomodulatory effects of MSC when compared to naive MSC [102, 116, 117]. Indeed, pretreating MSC (BM or placental) with IFNγ for 48 h suppressed T-cell proliferation to a greater extent than untreated MSC [113]. Cord-derived MSC had a greater suppressive effect than BMMSC when primed with IFNγ as assessed by T-cell proliferation assays and mixed lymphocyte reactions in vitro [102]. Furthermore, IL-2 secretion by T cells was significantly reduced when BMMSC, but not WJMSC, were primed with TNFα for 72 h prior to coculture in the presence of PHA [102]. However, enhancing MSC functions can have detri-

Table 5.2 Response of MSC to inflammatory environments

	Effect on MSC	MSC source	Species	Passage	References
Cytokine treatment					
IFNγ	↑ PD-L1, HGF and PGE$_2$ expression and IDO activity	BM/AD	Human/ mouse	2–10	[19, 102, 103]
	↓ TGFβ1 secretion	BM	Mouse	3–10	[103]
TNFα	↓ TGFβ1 and HGF secretion	BM/WJ	Human/ mouse	3–10	[102, 103]
	↑ TGFβ1 mRNA	WJ	Human	5–10	[94]
	↑ HGF, PGE$_2$ secretion	BM/WJ	Human/ mouse	3–8	[102, 103]
Poly(I:C)	↑ IDO, PGE$_2$, SMAD7 mRNA	BM	Human	≤4	
	↓ TGFβ1, IL-6, IL-8, CCL10, secretion	BM			[33]
	↑ Fibronectin deposition	–			
	↓ Differentiation capacity	–			
LPS	↑ Jagged-1/2, SMAD3 mRNA	BM	Human	≤4	[33, 94]
	↓ TGFβ1 and HGF expression	BM/AD			
	↑ Osteogenesis and collagen deposition	–			[33, 104]
	↓ Adipogenesis	–			
	↑ IL-1Ra, IL-6, IL-8, and IL-4 secretion	AD			
TGFβ1	↑ Migration	BM	Mouse	–	[91]
IFNγ+TNFα	↑ ICAM-1, VCAM-1, HIF-1α, VEGF, iNOS, PD-L1 expression	BM	Mouse	3–20	[18, 103, 105]
	↑ IL-6, IL-8, CXCL9, CXCL10 secretion	BM	Human/ mouse	–	[34, 106]
IL-1β+IFNγ +TNFα+IFNα	↑ IL-1β mRNA and IL-6 and IL-8 secretion	BM/WJ/ AD	Human	<2	[104]
	↑ TLR2, TLR3, ↓ TLR6 mRNA	BM/WJ/ AD			
	↑ TLR1 mRNA	WJ			
	↓ TLR5 mRNA	WJ/AD			
	↑ IFN-γ and ↓ HGF secretion	BM			
Disease					
RA	↓ MSC proliferation	BM	Human	1–6	[107, 108]
	Impaired ability to support haematopoiesis				
	↓ Cyclin-D; ↑ cyclin-D inhibitor				
SLE	↓ MSC proliferation	BM/WJ	Human/ mouse	>3	[109–112]
	↓ Differentiation into osteoblasts				

AD adipose, *BM* bone marrow, *DC* dendritic cell, *WJ* Wharton's jelly

Table 5.3 Effects of inflammatory cytokines on the immunomodulatory properties of MSC

	Effect	Mediator(s)	MSC source	Species	Passage	References
IFNγ	↓ Proliferation of T- and B-cells	IDO, PD-1	Placenta/BM/AD	Human/mouse	>2	[19, 73, 113]
	↓ B-cell differentiation into plasma cells	PD-1	BM	Mouse	20–25	[73]
	↓ Secretion of IFN-γ and TNFα by T-cell	–	BM	Human	≤10	[102]
	↓ Expansion of B$_{reg}$	IDO	AD	Human	2–5	[19]
TNFα	↓ DC maturation	–	BM	Mouse	3–10	[42]
	↓ CCR7 expression on DC					
	↓DC migration to CCL19					
	↓ Secretion of IFNγ and TNFα by T-cells	–	BM	Human	≤10	[113]
	↓ Splenocyte proliferation	PGE$_2$	BM	Mouse	3–10	[103]
IL-10	↓ T-cell proliferation	HLA-G5	BM	Human	1	[30]
	↓ NK cytotoxicity					
	↑ Expansion of T$_{reg}$					
IL-1β+IFNγ +TNFα+IFNα	↓ T-cell proliferation	–	BM/WJ/AD	Human	<2	[104]

AD adipose, *BM* bone marrow, *DC* dendritic cell, *IDO* indoleamine 2,3-dioxygenase, *PD1* programmed cell death 1, *PGE2* prostaglandin E2, *WJ* Wharton's jelly

mental effects. For example, IFNγ-stimulated MSC are better able to suppress B-cell proliferation but have a reduced capacity to induce B$_{reg}$ [19]. Co-injection of primed murine BMMSC (12 h TNFα and IFNγ) with a C26 colonic cancer cell line caused a significant increase in tumour growth when compared to untreated MSC [105]. That said, priming itself is not essential for the suppressive actions of MSC [17, 20, 21, 24–26, 38, 66]. But it does suggest that the MSC can respond to their local microenvironment, which in turn could affect their behaviour (reviewed by [118]). Whether priming of MSC in vitro is representative of the in vivo situation requires further research.

Engagement of TLR-3 in vitro was initially reported to enhance the effects of BMMSC, inducing the release of anti-inflammatory factors (e.g. IDO) [33]. In contrast, TLR-4 activation of BMMSC abrogated their ability to suppress T-lymphocyte proliferation and induced the release of pro-inflammatory cytokines (e.g. TNFα) and deposition collagen [33]. In vivo, systemic administration of TLR3-primed MSC ameliorated symptoms of lung injury and diabetic neuropathy, whilst TLR4-

primed MSC exacerbated disease compared to infusion of naive MSC [33]. Although MSC were defined as MSC2 and MSC1, respectively, it should be noted that these terms refer to the phenotype acquired following TLR activation, rather than the origin of the cells. Subsequent in vitro studies presented conflicting findings: IDO or PGE_2 secretion and T-cell proliferation have been reported to be enhanced, reduced, or unchanged by TLR 3- and TLR 4-stimulated BMMSC [34, 104, 119]. Different experimental conditions (treatment concentrations and duration) and the number and source of MSC are the likely explanation for these contradictory outcomes. Furthermore, MSC infusion has previously been shown to reduce lung oedema and inflammatory infiltrates in murine models of sepsis where high levels of LPS (TLR4 ligand) are present [49, 120]. As MSC dampened inflammation, rather than augment it, it is likely that the effects of TLR priming observed in vitro may not reflect MSC responses in vivo.

The behaviour of MSC is highly plastic, with the local inflammatory milieu (cytokines, danger signals, and bacterial components) having the potential to shape the immune regulatory effects of tissue-resident MSC [33]. Further work is essential to fully understand MSC biology during inflammatory responses and the impact of chronic inflammation. Such plasticity could have implications for MSC as a cell therapy – can we guarantee that the cells administered will maintain their immunosuppressive effects in a chronically inflamed site?

5.9 The Dangers of Chronic Inflammatory Environments on MSC Behaviour

Mesenchymal stromal cells *(see other chapters)*, including MSC, endogenously moderate inflammation, so why does it persist? Also, does chronic inflammation adversely and/or permanently affect MSC function? Ex vivo studies report that human BMMSC isolated from patients with RA have impaired ability to support haematopoiesis [107]. Furthermore, BMMSC from systemic lupus erythematosus (SLE) and RA patients have reduced proliferative capacity and reduced telomere length, indicative of a senescent phenotype when compared to healthy controls [108–110]. Likewise, reduced proliferation and osteogenesis were observed in BMMSC from patients with SLE and a murine preclinical model of SLE [111, 112]. In contrast, no such changes were observed in BMMSC isolated from patients with multiple sclerosis (MS; [121, 122]) or systemic sclerosis (SS; [35]). Importantly, MSC from patients with SLE, RA, and SS appear to maintain their immunomodulatory effector functions – as measured by T-cell proliferation assays [35, 108, 110]. Culturing healthy BMMSC in the presence of 20% synovial fluid from patients with osteoarthritis, but not post-mortem donors with no signs of joint inflammation, increased the gene expression of IL-6 and IDO [123]. Moreover, proteomic analysis of RA BMMSC revealed changes in molecules responsible for regulating cell cycle from G1 to S-phase when compared to healthy age and gender-matched controls, namely, an increase in cyclin-D inhibitors and decrease in cyclin-D [108]. The chronic inflammatory milieu appears to be capable of driving the proliferation and premature senescence of BMMSC, possibly contributing to further pathogenesis. Unfortunately, all of these studies analysed BMMSC, leaving the effect of the chronic inflammatory milieu on local tissue-resident MSC to be elucidated.

Ectopic fat deposits and/or alterations in local adipose tissue are associated with a number of disorders including Duchenne muscular dystrophy [124], myocardial infarction [125], type II diabetes [126], and RA [127–129]. Similarly aberrant bone formation or calcification has been described in fibrodysplasia ossificans progressiva [130], the vasculature of chronic kidney disease [131], and the adipose tissue in intra-abdominal surgery [132]. These deposits could be the result of inappropriate differentiation of tissue-resident MSC induced by inflammatory mediators in the affected tissue. Thus, under certain conditions, MSC could change their phenotype, no longer acting as brakes on the inflammatory response and possibly taking on a stimulatory state. This might occur during "classic" differentiation, e.g. into adipocytes, or conversion into a non-specific state in chronically insulted tissue. Indeed, MSC-derived adipocytes have lost the ability to suppress neutrophil capture to inflamed endothelium, as seen with undifferentiated MSC [133]. In a 3D multicellular migration assay, both MSC-derived adipocytes and osteoblasts were no longer able to suppress neutrophil adhesion to and migration through an inflamed endothelial monolayer, suggesting that transdifferentiation of MSC abrogates their immunomodulatory capacity [134]. In contrast, native stromal cells, adipocytes derived from them, and mature adipocytes from adipose tissue were all immuno-protective [133]. Thus disruption of normal tissue stroma homeostasis, as occurs in chronic inflammatory diseases, might drive "abnormal" adipogenesis which adversely influences the behaviour of MSC and contributes to pathogenic recruitment of leukocytes [133]. These novel findings parallel those we made when comparing stromal cells from healthy and diseased tissues, where stromal cells from chronically inflamed sites lost immunosuppressive properties and modified endothelial cells to inappropriately recruit leukocytes ([87, 88, 135–137]; see Chap. 3). Moreover, these effects were mediated by altering the bioactivity of IL-6 or TGFβ, making them act in a "pro-inflammatory" manner [138]. Whether a diseased environment (chronic inflammation or tumour) drives a similar pathogenic response in MSC remains to be addressed.

5.10 MSC in Therapeutics to Treat Inflammatory Disorders

The ability of MSC to modify immune responses has been the basis for clinical trials in a range of conditions [139]. Of these graft vs. host disease (GvHD) has been the most extensively studied, with early studies showing good therapeutic potential. To date there are 12 recently completed and 25 trials ongoing in this area; in all cases the outcomes have yet to be announced [139]. Systemic infusion of matched or mismatched BMMSC into patients with or at risk of GvHD improved clinical scores [140–142], with a few patients reporting complete remission at the 12-month follow-up [141, 143–145]. BMMSC therapy improved survival at 12–24 months in ~50–60% of patients with steroid-refractory GvHD in phase II trials [142, 146]. Due to the nature of these studies, none included a placebo control arm necessary to assess the true clinical benefit of MSC infusion. Early trials report therapeutic efficacy of MSC. However, randomised multicentre phase III trials of steroid-refractory GvHD showed no significant difference between treatment ("off-the-shelf" allogeneic MSC) and placebo groups [147]. The lack of efficacy may be due to differences in disease severity (degree of steroid resistance) between patients. Individuals with moderate disease severity may have a better response to MSC infusion compared to those with more severe disease, which could affect the outcome of trials. Recent follow-up studies have shown an increased incidence of haematological malignancies [148] or risk of pneumonia [149] in GvHD patients treated with MSC. That said, there was no evidence of tumour formations following intravenous infusion of MSC in patients with neuromyelitis optica spectrum disorder at 2 years follow-up [150]. The long-term risks and potential side effects of MSC therapy will need further investigation.

Based on promising data from preclinical models, trials are also examining the efficacy of MSC in autoimmune diseases, with a significant number involving patients with Crohn's disease, SLE, and RA (reviewed by [151]). However, a concern with these studies is the cyclical nature of patient's symptoms, making it difficult to determine whether improvements in the condition are due to the MSC or the natural disease cycle. As mentioned for GvHD, many of these studies also lack the appropriate placebo controls. Nevertheless, preclinical and clinical studies have shown potential clinical benefits of MSC treatment [53, 141, 143–145].

5.10.1 Limitations of Current Clinical Trials

Conflicting outcomes in clinical trials may arise from differences in trial design and lack of understanding of MSC biology. Variations in the clinical outcome of these trials may also be due to ex vivo expansion (passaging) of MSC which has a negative effect on their proliferation, differentiation, and immunosuppressive effects [152]. Due to the scarcity of MSC in tissues, large-scale culture ex vivo expansion is necessary to generate sufficient cell numbers for therapeutic administration, which may limit their clinical benefits. MSC are a heterogeneous population of cells with similar phenotypic features as other stromal populations such as fibroblasts. As such, MSC will need to be more stringently defined before becoming an "off-the-shelf" therapeutic strategy for treatment of inflammatory disorders.

Key concerns regarding the optimum route of administration, dose of MSC, the best source of cells, and the fate of the cells after infusion also need to be addressed (reviewed by [151]). Systemically infused MSC have a low homing efficiency (<1%) and become mechanically trapped in the lungs (reviewed by [153]), suggesting that the beneficial effects of MSC treatment are mostly likely due to soluble mediators [56]. However, a recent study reported that intravenously infused fluorescently labelled BMMSC initially lodged in the lungs but importantly were no longer detected at 24 h [153]. They subsequently have suggested that previous studies showing MSC redistribution in other tissues were detecting cell debris or phagocytosed MSC that are still labelled and postulated that any long-term immunosuppressive effects observed after MSC infusion are mediated by other cell types and not the MSC themselves [153]. For example, infused MSC can be phagocytosed by monocytes, inducing the monocytes to acquire the non-classical anti-inflammatory phenotype through up-regulation of CD16 and therefore transferring their MSC immunomodulatory effects onto the monocytes [154]. Alternatively, human BMMSC-derived apoptotic bodies have been suggested to initiate MSC-induced immunosuppressive in a murine model of GvHD [155].

The long-term effects of MSC treatment (5–10 years follow-up) have not been carried out. Any long-term risks of MSC treatment are currently unknown, and issues such as MSC response to other therapeutic interventions, potential tumorigenicity, and tissue distribution upon administration will need to be addressed to eliminate possible risks of MSC treatment. Manipulating the functions of endogenous MSC for therapeutic use may therefore be an attractive alternative to current treatment modalities for inflammatory conditions. However, without fully ascertaining the mode of actions of endogenous MSC, it will be difficult to elucidate their true therapeutic potential.

5.11   Conclusions

Tissue-resident MSC are endogenous regulators of inflammation. They have an inherent capacity to sense even the subtlest of changes in their microenvironment and respond accordingly. MSC maintain tissue homeostasis: replacing damaged cells through their differentiation into the target stromal cell and also supporting the haematopoietic niche. During inflammation, MSC inhibit the archetypical inflammatory behaviours of their target cell whilst simultaneously promoting anti-inflammatory, pro-resolution agents and/or the generation of regulatory cells. We, ourselves, have shown that MSC communicate with blood vascular endothelial cells to regulate the inflammatory infiltrate. MSC predominately mediate their effects through the release of soluble factors, but in certain context, direct cell-cell interactions are thought to be required to enhance these further. Whilst MSC cell therapy is currently being explored for clinical benefit, many of the clinical trials are in the earliest phases with the outcomes yet to be announced or inconclusive. Such studies are confounded by differences in their design, source, and dose of MSC and the absence of placebo controls, making it difficult to ascertain the true clinical benefit of MSC treatment. Further research is needed to understand how MSC communicate with cells, other than leukocytes, within tissues and whether these interactions change during an inflammatory response. Moreover, it is critical we understand the impact chronic inflammation has on the function of MSC. Can we guarantee that therapeutic MSC will maintain their immunosuppressive effects in a chronically inflamed site? Whether MSC-derived media or effector molecules (either from

MSC or cocultures with other cell types) would be a safer and more efficacious alternative intervention remains to be seen.

Conflicts of Interest The authors declare that they have no conflicts of interest.

Funding HM and LSCW were supported by BBSRC and MRC PhD studentships, respectively. HMM was supported by an Arthritis Research UK Career Development Fellowship (19899) and Systems Science for Health, University of Birmingham (5212).

References

1. Pal R, Hanwate M, Jan M, Totey S. Phenotypic and functional comparison of optimum culture conditions for upscaling of bone marrow-derived mesenchymal stem cells. J Tissue Eng Regen Med. 2009;3:163–74.
2. Sotiropoulou PA, Perez SA, Salagianni M, Baxevanis CN, Papamichail M. Characterization of the optimal culture conditions for clinical scale production of human mesenchymal stem cells. Stem Cells. 2006;24(2):462–71.
3. Christodoulou I, Kolisis FN, Papaevangeliou D, Zoumpourlis V. Comparative evaluation of human mesenchymal stem cells of fetal (Wharton's jelly) and adult (adipose tissue) origin during prolonged in vitro expansion: considerations for Cytotherapy. Stem Cells Int. 2013;2013:246134.
4. Baksh D, Yao R, Tuan RS. Comparison of proliferative and multilineage differentiation potential of human mesenchymal stem cells derived from umbilical cord and bone marrow. Stem Cells. 2007;25:1384–92.
5. Jin H, Bae Y, Kim M, Kwon S-J, Jeon H, Choi S, et al. Comparative analysis of human mesenchymal stem cells from bone marrow, adipose tissue, and umbilical cord blood as sources of cell therapy. Int J Mol Sci. 2013;14:17986–8001.
6. Kern S, Eichler H, Stoeve J, Klüter H, Bieback K. Comparative analysis of mesenchymal stem cells from bone marrow, umbilical cord blood, or adipose tissue. Stem Cells. 2006;24:1294–301.
7. Magatti M, De Munari S, Vertua E, Gibelli L, Wengler GS, Parolini O. Human amnion mesenchyme harbors cells with allogeneic T-cell suppression and stimulation capabilities. Stem Cells. 2008;26:182–92.
8. Wolbank S, Peterbauer A, Fahrner M, Hennerbichler S, van Griensven M, Stadler G, et al. Dose-dependent immunomodulatory effect of human stem cells from amniotic membrane: a comparison with human mesenchymal stem cells from adipose tissue. Tissue Eng. 2007;13:1173–83.
9. Chen P-M, Yen M-L, Liu K-J, Sytwu H-K, Yen B-L. Immunomodulatory properties of human adult and fetal multipotent mesenchymal stem cells. J Biomed Sci. 2011;18(1):49.
10. Chang C-J, Yen M-L, Chen Y-C, Chien C-C, Huang H-I, Bai C-H, et al. Placenta-derived multipotent cells exhibit immunosuppressive properties that are enhanced in the presence of interferon-gamma. Stem Cells. 2006;24:2466–77.
11. Méndez-Ferrer S, Michurina TV, Ferraro F, Mazloom AR, Macarthur BD, Lira SA, et al. Mesenchymal and haematopoietic stem cells form a unique bone marrow niche. Nature. 2010;466(7308):829–34.
12. Sugiyama T, Kohara H, Noda M, Nagasawa T. Maintenance of the hematopoietic stem cell pool by CXCL12-CXCR4 chemokine signaling in bone marrow stromal cell niches. Immunity. 2006;25(6):977–88.

13. Chow A, Lucas D, Hidalgo A, Méndez-Ferrer S, Hashimoto D, Scheiermann C, et al. Bone marrow CD169+ macrophages promote the retention of hematopoietic stem and progenitor cells in the mesenchymal stem cell niche. J Exp Med. 2011;208(2):261–71.
14. Ahn JY, Park G, Shim JS, Lee JW, Oh IH. Intramarrow injection of beta-catenin-activated, but not naive mesenchymal stromal cells stimulates self-renewal of hematopoietic stem cells in bone marrow. Exp Mol Med. 2010;42(2):122–31.
15. Omatsu Y, Sugiyama T, Kohara H, Kondoh G, Fujii N, Kohno K, et al. The essential functions of adipo-osteogenic progenitors as the hematopoietic stem and progenitor cell niche. Immunity. 2010;33(3):387–99.
16. Brandau S, Jakob M, Hemeda H, Bruderek K, Janeschik S, Bootz F, et al. Tissue-resident mesenchymal stem cells attract peripheral blood neutrophils and enhance their inflammatory activity in response to microbial challenge. J Leukoc Biol. 2010;88(5):1005–15.
17. Raffaghello L, Bianchi G, Bertolotto M, Montecucco F, Busca A, Dallegri F, et al. Human mesenchymal stem cells inhibit neutrophil apoptosis: a model for neutrophil preservation in the bone marrow niche. Stem Cells. 2008;26:151–62.
18. Ren G, Zhao X, Zhang L, Zhang J, L'Huillier A, Ling W, et al. Inflammatory cytokine-induced intercellular adhesion molecule-1 and vascular cell adhesion molecule-1 in mesenchymal stem cells are critical for immunosuppression. J Immunol. 2010;184(5):2321–8.
19. Luk F, Carreras-Planella L, Korevaar SS, de Witte SFH, Borràs FE, Betjes MGH, et al. Inflammatory conditions dictate the effect of mesenchymal stem or stromal cells on B cell function. Front Immunol. 2017;8:1042.
20. Le Blanc K, Mougiakakos D. Multipotent mesenchymal stromal cells and the innate immune system. Nat Rev Immunol. 2012;12:383–96.
21. Spaggiari GM, Capobianco A, Abdelrazik H, Becchetti F, Mingari MC, Moretta L. Mesenchymal stem cells inhibit natural killer-cell proliferation, cytotoxicity, and cytokine production: role of indoleamine 2,3-dioxygenase and prostaglandin E2. Blood. 2008;111(3):1327–33.
22. Jiang X-X, Zhang Y, Liu B, Zhang S-X, Wu Y, Yu X-D, et al. Human mesenchymal stem cells inhibit differentiation and function of monocyte-derived dendritic cells. Blood. 2005;105(10):4120–6.
23. Shi C, Jia T, Mendez-Ferrer S, Hohl TM, Serbina NV, Lipuma L, et al. Bone marrow mesenchymal stem and progenitor cells induce monocyte emigration in response to circulating toll-like receptor ligands. Immunity. 2011;34(4):590–601.
24. Kim J, Hematti P. Mesenchymal stem cell-educated macrophages: a novel type of alternatively activated macrophages. Exp Hematol. 2009;37:1445–53.
25. Maggini J, Mirkin G, Bognanni I, Holmberg J, Piazzón IM, Nepomnaschy I, et al. Mouse bone marrow-derived mesenchymal stromal cells turn activated macrophages into a regulatory-like profile. PLoS One. 2010;5(2):e9252.
26. François M, Romieu-Mourez R, Li M, Galipeau J. Human MSC suppression correlates with cytokine induction of Indoleamine 2,3-dioxygenase and bystander M2 macrophage differentiation. Mol Ther. 2012;20:187–95.
27. Ma S, Xie N, Li W, Yuan B, Shi Y, Wang Y. Immunobiology of mesenchymal stem cells. Cell Death Differ. 2014;21(2):216–25.
28. Benvenuto F, Ferrari S, Gerdoni E, Gualandi F, Frassoni F, Pistoia V, et al. Human mesenchymal stem cells promote survival of T cells in a quiescent state. Stem Cells. 2007;25:1753–60.
29. Glennie S, Soeiro I, Dyson PJ, Lam EW-F, Dazzi F. Bone marrow mesenchymal stem cells induce division arrest anergy of activated T cells. Blood. 2005;105:2821–7.
30. Selmani Z, Naji A, Zidi I, Favier B, Gaiffe E, Obert L, et al. Human leukocyte antigen-G5 secretion by human mesenchymal stem cells is required to suppress T lymphocyte and natural killer function and to induce CD4+CD25highFOXP3+ regulatory T cells. Stem Cells. 2008;26(1):212–22.
31. Gu YZ, Xue Q, Chen YJ, Yu GH, de Qing M, Shen Y, et al. Different roles of PD-L1 and FasL in immunomodulation mediated by human placenta-derived mesenchymal stem cells. Hum Immunol. 2013;74(3):267–76.

32. Augello A, Tasso R, Negrini SM, Amateis A, Indiveri F, Cancedda R, et al. Bone marrow mesenchymal progenitor cells inhibit lymphocyte proliferation by activation of the programmed death 1 pathway. Eur J Immunol. 2005;35(5):1482–90.
33. Waterman RS, Tomchuck SL, Henkle SL, Betancourt AM. A new mesenchymal stem cell (MSC) paradigm: polarization into a pro-inflammatory MSC1 or an immunosuppressive MSC2 phenotype. PLoS One. 2010;5(4):e10088.
34. Liotta F, Angeli R, Cosmi L, Filì L, Manuelli C, Frosali F, et al. Toll-like receptors 3 and 4 are expressed by human bone marrow-derived mesenchymal stem cells and can inhibit their T-cell modulatory activity by impairing Notch signaling. Stem Cells. 2008;26:279–89.
35. Larghero J, Farge D, Braccini A, Lecourt S, Scherberich A, Fois E, et al. Phenotypical and functional characteristics of in vitro expanded bone marrow mesenchymal stem cells from patients with systemic sclerosis. Ann Rheum Dis. 2008;67(4):443–9.
36. Di Nicola M, Carlo-Stella C, Magni M, Milanesi M, Longoni PD, Matteucci P, et al. Human bone marrow stromal cells suppress T-lymphocyte proliferation induced by cellular or non-specific mitogenic stimuli. Blood. 2002;99(10):3838–43.
37. Sato K, Ozaki K, Oh I, Meguro A, Hatanaka K, Nagai T, et al. Nitric oxide plays a critical role in suppression of T-cell proliferation by mesenchymal stem cells. Blood. 2007;109(1):228–34.
38. Aggarwal S, Pittenger MF. Human mesenchymal stem cells modulate allogeneic immune cell responses. Blood. 2005;105:1815–22.
39. Beyth S, Borovsky Z, Mevorach D, Liebergall M, Gazit Z, Aslan H, et al. Human mesenchymal stem cells alter antigen-presenting cell maturation and induce T-cell unresponsiveness. Blood. 2005;105(5):2214–9.
40. Corcione A, Benvenuto F, Ferretti E, Giunti D, Cappiello V, Cazzanti F, et al. Human mesenchymal stem cells modulate B-cell functions. Blood. 2006;107(1):367–72.
41. Ramasamy R, Fazekasova H, Lam EW-F, Soeiro I, Lombardi G, Dazzi F. Mesenchymal stem cells inhibit dendritic cell differentiation and function by preventing entry into the cell cycle. Transplantation. 2007;83(1):71–6.
42. English K, Barry FP, Mahon BP. Murine mesenchymal stem cells suppress dendritic cell migration, maturation and antigen presentation. Immunol Lett. 2008;115(1):50–8.
43. Rahbarghazi R, Nassiri SM, Khazraiinia P, Kajbafzadeh A, Ahmadi SH, Mohammadi E, et al. Juxtacrine and paracrine interactions of rat marrow derived mesenchymal stem cells, muscle derived satellite cells and neonatal cardiomyocytes with endothelial cells in angiogenesis dynamics. Stem Cells Dev. 2013;22(6):855–65.
44. Chen L, Tredget EE, Wu PYG, Wu Y, Wu Y. Paracrine factors of mesenchymal stem cells recruit macrophages and endothelial lineage cells and enhance wound healing. PLoS One. 2008;3(4):e1886.
45. Dhar K, Dhar G, Majumder M, Haque I, Mehta S, Van Veldhuizen PJ, et al. Tumor cell-derived PDGF-B potentiates mouse mesenchymal stem cells-pericytes transition and recruitment through an interaction with NRP-1. Mol Cancer. 2010;9:209.
46. Otsu K, Das S, Houser SD, Quadri SK, Bhattacharya S, Bhattacharya J. Concentration-dependent inhibition of angiogenesis by mesenchymal stem cells. Hematop Stem Cells. 2009;113(18):4197–205.
47. Pati S, Gerber MH, Menge TD, Wataha KA, Zhao Y, Baumgartner JA, et al. Bone marrow derived mesenchymal stem cells inhibit inflammation and preserve vascular endothelial integrity in the lungs after hemorrhagic shock. PLoS One. 2011;6(9):e25171.
48. Zhao YD, Ohkawara H, Vogel SM, Malik AB, Zhao Y-Y. Bone marrow-derived progenitor cells prevent thrombin-induced increase in lung vascular permeability. Am J Physiol Lung Cell Mol Physiol. 2010;298(1):L36–44.
49. Zhao YD, Ohkawara H, Rehman J, Wary KK, Vogel SM, Minshall RD, et al. Bone marrow progenitor cells induce endothelial adherens junction integrity by sphingosine-1-phosphate-mediated Rac1 and Cdc42 signaling. Circ Res. 2009;105(7):696–704.

50. Pati S, Khakoo AY, Zhao J, Jimenez F, Gerber MH, Harting M, et al. Human mesenchymal stem cells inhibit vascular permeability by modulating vascular endothelial cadherin/b-catenin signaling. Stem Cells Dev. 2010;20(1):89–101.
51. Luu NT, McGettrick HM, Buckley CD, Newsome P, Ed Rainger G, Frampton J, et al. Crosstalk between mesenchymal stem cells and endothelial cells leads to down-regulation of cytokine-iNduced leukocyte recruitment. Stem Cells. 2013;31:2690–702.
52. Munir H, Rainger GE, Nash GBMH, McGettrick H. Analyzing the effects of stromal cells on the recruitment of leukocytes from flow. J Vis Exp. 2015;95:e52480.
53. Munir H, Luu N-T, Clarke LSC, Nash GB, McGettrick HM. Comparative ability of mesenchymal stromal cells from different tissues to limit neutrophil recruitment to inflamed endothelium. PLoS One. 2016;11(5):e0155161.
54. Dominici M, Le Blanc K, Mueller I, Slaper-Cortenbach I, Marini F, Krause D, et al. Minimal criteria for defining multipotent mesenchymal stromal cells. The International Society for Cellular Therapy position statement. Cytotherapy. 2006;8:315–7.
55. Tormin A, Li O, Brune JC, Walsh S, Schütz B, Ehinger M, et al. CD146 expression on primary nonhematopoietic bone marrow stem cells is correlated with in situ localization. Blood. 2011;117:5067–77.
56. Eggenhofer E, Luk F, Dahlke MH, Hoogduijn MJ. The life and fate of mesenchymal stem cells. Front Immunol. 2014;5:148.
57. Morikawa S, Mabuchi Y, Niibe K, Suzuki S, Nagoshi N, Sunabori T, et al. Development of mesenchymal stem cells partially originate from the neural crest. Biochem Biophys Res Commun. 2009;379(4):1114–9.
58. Nagoshi N, Shibata S, Kubota Y, Nakamura M, Nagai Y, Satoh E, et al. Ontogeny and multipotency of neural crest-derived stem cells in mouse bone marrow, dorsal root ganglia, and whisker pad. Cell Stem Cell. 2008;2(4):392–403.
59. Crisan M, Yap S, Casteilla L, Chen C-W, Corselli M, Park TS, et al. A perivascular origin for mesenchymal stem cells in multiple human organs. Cell Stem Cell. 2008;3(3):301–13.
60. Feng J, Mantesso A, De Bari C, Nishiyama A, Sharpe PT. Dual origin of mesenchymal stem cells contributing to organ growth and repair. Proc Natl Acad Sci U S A. 2011;108(16):6503–8.
61. Vodyanik MA, Yu J, Zhang X, Tian S, Stewart R, Thomson JA, et al. A mesoderm-derived precursor for mesenchymal stem and endothelial cells. Cell Stem Cell. 2010;7(6):718–29.
62. Li N, Hua J. Interactions between mesenchymal stem cells and the immune system. Cell Mol Life Sci. 2017;74(13):2345–60.
63. Brandau S, Jakob M, Bruderek K, Bootz F, Giebel B, Radtke S, et al. Mesenchymal stem cells augment the anti-bacterial activity of neutrophil granulocytes. Jacobs R, editor. PLoS One. 2014;9(9):e106903.
64. Zhu Q, Zhang X, Zhang L, Li W, Wu H, Yuan X, et al. The IL-6-STAT3 axis mediates a reciprocal crosstalk between cancer-derived mesenchymal stem cells and neutrophils to synergistically prompt gastric cancer progression. Cell Death Dis. 2014;5:e1295.
65. Khan I, Zhang L, Mohammed M, Archer FE, Abukharmah J, Yuan Z, et al. Effects of Wharton's jelly-derived mesenchymal stem cells on neonatal neutrophils. J Inflamm Res. 2014;8:1–8.
66. Sotiropoulou PA, Perez SA, Gritzapis AD, Baxevanis CN, Papamichail M. Interactions between human mesenchymal stem cells and natural killer cells. Stem Cells. 2006;24:74–85.
67. Németh K, Leelahavanichkul A, Yuen PST, Mayer B, Parmelee A, Doi K, et al. Bone marrow stromal cells attenuate sepsis via prostaglandin E(2)-dependent reprogramming of host macrophages to increase their interleukin-10 production. Nat Med. 2009;15:42–9.
68. Letourneau PA, Menge TD, Wataha KA, Wade CE, Cox SC Jr, Holcomb JB, et al. Human bone marrow derived mesenchymal stem cells regulate leukocyte-endothelial interactions and activation of transcription factor. J Tissue Sci Eng. 2011;3:1–7.
69. Bernardo ME, Fibbe WE. Mesenchymal stromal cells: sensors and switchers of inflammation. Cell Stem Cell. 2013;13:392–402.

70. Rubtsov Y, Goryunov K, Romanov A, Suzdaltseva Y, Sharonov G, Tkachuk V. Molecular mechanisms of immunomodulation properties of mesenchymal stromal cells: a new insight into the role of ICAM-1. Stem Cells Int. 2017;2017:6516854.
71. Franquesa M, Mensah FK, Huizinga R, Strini T, Boon L, Lombardo E, et al. Human adipose tissue-derived mesenchymal stem cells abrogate plasmablast formation and induce regulatory B cells independently of T helper cells. Stem Cells. 2015;33(3):880–91.
72. Blair PA, Noreña LY, Flores-Borja F, Rawlings DJ, Isenberg DA, Ehrenstein MR, et al. CD19+CD24hiCD38hi B cells exhibit regulatory capacity in healthy individuals but are functionally impaired in systemic lupus erythematosus patients. Immunity. 2010;32(1):129–40.
73. Schena F, Gambini C, Gregorio A, Mosconi M, Reverberi D, Gattorno M, et al. Interferon-γ–dependent inhibition of B cell activation by bone marrow–derived mesenchymal stem cells in a murine model of systemic lupus erythematosus. Arthritis Rheum. 2010;62(9):2776–86.
74. Wu J, Ji C, Cao F, Lui H, Xia B, Wang L. Bone marrow mesenchymal stem cells inhibit dendritic cells differentiation and maturation by microRNA-23b. Biosci Rep. 2017;37(2):BSR20160436.
75. Detry O, Vandermeulen M, Delbouille M-H, Somja J, Bletard N, Briquet A, et al. Infusion of mesenchymal stromal cells after deceased liver transplantation: a phase I–II, open-label, clinical study. J Hepatol. 2017;67(1):47–55.
76. Sheriff L, Alanazi A, Ward LSC, Ward C, Munir H, Rayes J, et al. Origin-specific adhesive interactions of mesenchymal stem cells with platelets influence their behaviour after infusion. Stem Cells. 2018; https://doi.org/10.1002/stem.2811. [Epub ahead of print]
77. Langer HF, Stellos K, Steingen C, Froihofer A, Schönberger T, Krämer B, et al. Platelet derived bFGF mediates vascular integrative mechanisms of mesenchymal stem cells in vitro. J Mol Cell Cardiol. 2009;47(2):315–25.
78. Jiang L, Song XH, Liu P, Zeng CL, Huang ZS, Zhu LJ, et al. Platelet-mediated mesenchymal stem cells homing to the lung reduces monocrotaline-induced rat pulmonary hypertension. Cell Transplant. 2012;21(7):1463–75.
79. Teo GSL, Yang Z, Carman CV, Karp JM, Lin CP. Intravital imaging of mesenchymal stem cell trafficking and association with platelets and neutrophils. Stem Cells. 2015;33(1):265–77.
80. Wu Y, Chen L, Scott PG, Tredget EE. Mesenchymal stem cells enhance wound healing through differentiation and angiogenesis. Stem Cells. 2007;25(10):2648–59.
81. Nassiri SM, Rahbarghazi R. Interactions of mesenchymal stem cells with endothelial cells. Stem Cells Dev. 2014;23(4):319–32.
82. Suzuki K, Sun R, Origuchi M, Kanehira M, Takahata T, Itoh J, et al. Mesenchymal stromal cells promote tumor growth through the enhancement of neovascularization. Mol Med. 2011;17(7–8):579–87.
83. Anderson JD, Johansson HJ, Graham CS, Vesterlund M, Pham MT, Bramlett CS, et al. Comprehensive proteomic analysis of mesenchymal stem cell exosomes reveals modulation of angiogenesis via nuclear factor-kappaB signaling. Stem Cells. 2016;34(3):601–13.
84. Potter DR, Miyazawa BY, Gibb SL, Deng X, Togaratti PP, Croze RH, et al. Mesenchymal stem cell-derived extracellular vesicles attenuate pulmonary vascular permeability and lung injury induced by hemorrhagic shock and trauma. J Trauma Acute Care Surg. 2018;84(2):245–56.
85. Sinha S, Iyer D, Granata A. Embryonic origins of human vascular smooth muscle cells: implications for in vitro modeling and clinical application. Cell Mol Life Sci. 2014;71(12):2271–88.
86. Kuravi SJ, McGettrick HM, Satchell SC, Saleem MA, Harper L, Williams JM, et al. Podocytes regulate neutrophil recruitment by glomerular endothelial cells via IL-6-mediated crosstalk. J Immunol. 2014;193(1):234–43.
87. Lally F, Smith E, Filer A, Stone MA, Shaw JS, Nash GB, et al. A novel mechanism of neutrophil recruitment in a coculture model of the rheumatoid synovium. Arthritis Rheum. 2005;52(11):3460–9.
88. McGettrick HM, Smith E, Filer A, Kissane S, Salmon M, Buckley CD, et al. Fibroblasts from different sites may promote or inhibit recruitment of flowing lymphocytes by endothelial cells. Eur J Immunol. 2009;39:113–25.

89. Aird WC. Endothelial cell heterogeneity. Cold Spring Harb Perspect Med. 2012;2(1):a006429.
90. Parsonage G, Filer AD, Haworth O, Nash GB, Rainger GE, Salmon M, et al. A stromal address code defined by fibroblasts. Trends Immunol. 2005;26(3):150–6.
91. Tang Y, Wu X, Lei W, Pang L, Wan C, Shi Z, et al. TGF-beta1-induced migration of bone mesenchymal stem cells couples bone resorption with formation. Nat Med. 2009;15(7):757–65.
92. Abdel Aziz MT, Atta HM, Mahfouz S, Fouad HH, Roshdy NK, Ahmed HH, et al. Therapeutic potential of bone marrow-derived mesenchymal stem cells on experimental liver fibrosis. Clin Biochem. 2007;40(12):893–9.
93. Semedo P, Correa-Costa M, Cenedeze MA, Malheiros DMAC, Dos Reis MA, Shimizu MH, et al. Mesenchymal stem cells attenuate renal fibrosis through immune modulation and remodeling properties in a rat remnant kidney model. Stem Cells. 2009;27(12):3063–73.
94. Wu Y, Peng Y, Gao D, Feng C, Yuan X, Li H, et al. Mesenchymal stem cells suppress fibroblast proliferation and reduce skin fibrosis through a TGF-β3-dependent activation. Int J Low Extrem Wounds. 2015;14(1):50–62.
95. Li L, Zhang Y, Li Y, Yu B, Xu Y, Zhao S, et al. Mesenchymal stem cell transplantation attenuates cardiac fibrosis associated with isoproterenol-induced global heart failure. Transpl Int. 2008;21(12):1181–9.
96. Park SJ, Kim KJ, Kim WU, Cho CS. Interaction of mesenchymal stem cells with fibroblast-like synoviocytes via cadherin-11 promotes the angiogenesis by enhanced secretion of placental growth factor. J Immunol. 2014;192(7):3003–10.
97. Yates CC, Rodrigues M, Nuschke A, Johnson ZI, Whaley D, Stolz D, et al. Multipotent stromal cells/mesenchymal stem cells and fibroblasts combine to minimize skin hypertrophic scarring. Stem Cell Res Ther. 2017;8(1):193.
98. Dalby MJ, Gadegaard N, Oreffo ROC. Harnessing nanotopography and integrin-matrix interactions to influence stem cell fate. Nat Mater. 2014;13(6):558–69.
99. Lee J, Abdeen AA, Kilian KA. Rewiring mesenchymal stem cell lineage specification by switching the biophysical microenvironment. Sci Rep. 2014;4:5188.
100. McMurray RJ, Gadegaard N, Tsimbouri PM, Burgess KV, McNamara LE, Tare R, et al. Nanoscale surfaces for the long-term maintenance of mesenchymal stem cell phenotype and multipotency. Nat Mater. 2011;10(8):637–44.
101. Dalby MJ, Gadegaard N, Tare R, Andar A, Riehle MO, Herzyk P, et al. The control of human mesenchymal cell differentiation using nanoscale symmetry and disorder. Nat Mater. 2007;6(12):997–1003.
102. Prasanna SJ, Gopalakrishnan D, Shankar SR, Vasandan AB. Pro-inflammatory cytokines, IFNgamma and TNFalpha, influence immune properties of human bone marrow and Wharton jelly mesenchymal stem cells differentially. PLoS One. 2010;5:e9016.
103. English K, Barry FP, Field-Corbett CP, Mahon BP. IFN-gamma and TNF-alpha differentially regulate immunomodulation by murine mesenchymal stem cells. Immunol Lett. 2007;110:91–100.
104. Raicevic G, Najar M, Stamatopoulos B, De Bruyn C, Meuleman N, Bron D, et al. The source of human mesenchymal stromal cells influences their TLR profile as well as their functional properties. Cell Immunol. 2011;270:207–16.
105. Liu Y, Han ZP, Zhang SS, Jing YY, Bu XX, Wang CY, et al. Effects of inflammatory factors on mesenchymal stem cells and their role in the promotion of tumor angiogenesis in colon cancer. J Biol Chem. 2011;286(28):25007–15.
106. Ren G, Zhang L, Zhao X, Xu G, Zhang Y, Roberts AI, et al. Mesenchymal stem cell-mediated immunosuppression occurs via concerted action of chemokines and nitric oxide. Cell Stem Cell. 2008;2(2):141–50.
107. Papadaki HA, Kritikos HD, Gemetzi C, Koutala H, JCW M, Boumpas DT, et al. Bone marrow progenitor cell reserve and function and stromal cell function are defective in rheumatoid arthritis: evidence for a tumor necrosis factor alpha-mediated effect. Blood. 2002;99(5):1610–9.

108. Kastrinaki M-C, Sidiropoulos P, Roche S, Ringe J, Lehmann S, Kritikos H, et al. Functional, molecular and proteomic characterisation of bone marrow mesenchymal stem cells in rheumatoid arthritis. Ann Rheum Dis. 2008;67:741–9.

109. Sun L, Wang D, Liang J, Zhang H, Feng X, Wang H, et al. Umbilical cord mesenchymal stem cell transplantation in severe and refractory systemic lupus erythematosus. Arthritis Rheum. 2010;62:2467–75.

110. Nie Y, Lau C, Lie A, Chan G, Mok M. Defective phenotype of mesenchymal stem cells in patients with systemic lupus erythematosus. Lupus. 2010;19(7):850–9.

111. El-Badri NS, Hakki A, Ferrari A, Shamekh R, Good RA. Autoimmune disease: is it a disorder of the microenvironment? Immunol Res. 2008;41(1):79–86.

112. Tang Y, Xie H, Chen J, Geng L, Chen H, Li X, et al. Activated NF-kappaB in bone marrow mesenchymal stem cells from systemic lupus erythematosus patients inhibits osteogenic differentiation through downregulating Smad signaling. Stem Cells Dev. 2013;22(4):668–78.

113. Jones BJ, Brooke G, Atkinson K, McTaggart SJ. Immunosuppression by placental Indoleamine 2,3-dioxygenase: a role for mesenchymal stem cells. Placenta. 2007;28(11–12):1174–81.

114. Bahra P, Rainger GE, Wautier JL, Nguyet-Thin L, Nash GB. Each step during transendothelial migration of flowing neutrophils is regulated by the stimulatory concentration of tumour necrosis factor-alpha. Cell Adhes Commun. 1998;6:491–501.

115. Filer A, Parsonage G, Smith E, Osborne C, Thomas AMC, Curnow SJ, et al. Differential survival of leukocyte subsets mediated by synovial, bone marrow, and skin fibroblasts: site-specific versus activation-dependent survival of T cells and neutrophils. Arthritis Rheum. 2006;54(7):2096–108.

116. Deuse T, Stubbendorff M, Tang-Quan K, Phillips N, Kay MA, Eiermann T, et al. Immunogenicity and immunomodulatory properties of umbilical cord lining mesenchymal stem cells. Cell Transplant. 2011;20:655–67.

117. Yoo KH, Jang IK, Lee MW, Kim HE, Yang MS, Eom Y, et al. Comparison of immunomodulatory properties of mesenchymal stem cells derived from adult human tissues. Cell Immunol. 2009;259:150–6.

118. Najar M, Krayem M, Meuleman N, Bron D, Lagneaux L. Mesenchymal stromal cells and toll-like receptor priming: a critical review. Immune Netw. 2017;17(2):89–102.

119. Opitz CA, Litzenburger UM, Lutz C, Lanz TV, Tritschler I, Köppel A, et al. Toll-like receptor engagement enhances the immunosuppressive properties of human bone marrow-derived mesenchymal stem cells by inducing indoleamine-2,3-dioxygenase-1 via interferon-beta and protein kinase R. Stem Cells. 2009;27:909–19.

120. Tyndall A, Pistoia V. Mesenchymal stem cells combat sepsis. Nat Med. 2009;15:18–20.

121. Mallam E, Kemp K, Wilkins A, Rice C, Scolding N. Characterization of in vitro expanded bone marrow-derived mesenchymal stem cells from patients with multiple sclerosis. Mult Scler. 2010;16(8):909–18.

122. Papadaki HA, Tsagournisakis M, Mastorodemos V, Pontikoglou C, Damianaki A, Pyrovolaki K, et al. Normal bone marrow hematopoietic stem cell reserves and normal stromal cell function support the use of autologous stem cell transplantation in patients with multiple sclerosis. Bone Marrow Transplant. 2005;36(12):1053–63.

123. Leijs MJC, van Buul GM, Lubberts E, Bos PK, Verhaar JAN, Hoogduijn MJ, et al. Effect of arthritic synovial fluids on the expression of immunomodulatory factors by mesenchymal stem cells: an explorative in vitro study. Front Immunol. 2012;3:231.

124. Uezumi A, Fukada S, Yamamoto N, Takeda S, Tsuchida K. Mesenchymal progenitors distinct from satellite cells contribute to ectopic fat cell formation in skeletal muscle. Nat Cell Biol. 2010;12(2):143–52.

125. Goldfarb JW, Roth M, Han J. Myocardial fat deposition after left ventricular myocardial infarction: assessment by using MR water-fat separation imaging. Radiology. 2009;253(1):65–73.

126. Goodpaster BH, Wolf D. Skeletal muscle lipid accumulation in obesity, insulin resistance, and type 2 diabetes. Pediatr Diabetes. 2004;5(4):219–26.

127. Arend WP, Mehta G, Antonioli AH, Takahashi M, Takahashi K, Stahl GL, et al. Roles of adipocytes and fibroblasts in activation of the alternative pathway of complement in inflammatory arthritis in mice. J Immunol. 2013;190(12):6423–33.
128. Clements KM, Ball AD, Jones HB, Brinckmann S, Read SJ, Murray F. Cellular and histopathological changes in the infrapatellar fat pad in the monoiodoacetate model of osteoarthritis pain. Osteoarthr Cartil. 2009;17(6):805–12.
129. Schweitzer ME, Falk a, Pathria M, Brahme S, Hodler J, Resnick D. MR imaging of the knee: can changes in the intracapsular fat pads be used as a sign of synovial proliferation in the presence of an effusion? Am J Roentgenol. 1993;160(4):823–6.
130. Culbert AL, Chakkalakal SA, Theosmy EG, Brennan TA, Kaplan FS, Shore EM. Alk2 regulates early chondrogenic fate in fibrodysplasia ossificans progressiva heterotopic endochondral ossification. Stem Cells. 2014;32(5):1289–300.
131. Mizobuchi M, Towler D, Slatopolsky E. Vascular calcification: the killer of patients with chronic kidney disease. J Am Soc Nephrol. 2009;20(7):1453–64.
132. Zamolyi RQ, Souza P, Nascimento AG, Unni KK. Intraabdominal myositis ossificans: a report of 9 new cases. Int J Surg Pathol. 2006;14:37–41.
133. Munir H, Ward LSC, Sheriff L, Kemble S, Nayar S, Barone F, et al. Adipogenic differentiation of mesenchymal stem cells alters their immunomodulatory properties in a tissue-specific manner. Stem Cells. 2017;35(6):1636–46.
134. Munir H. *Mesenchymal stem cells as endogenous regulators of leukocyte recruitment; the effects of differentiation.* [PhD thesis on the Internet]. Birmingham: University of Birmingham; 2016 [cited 2018 Feb 28]. Available from: http://etheses.bham.ac.uk/.
135. McGettrick HM, Buckley CD, Filer A, Rainger GE, Nash GB. Stromal cells differentially regulate neutrophil and lymphocyte recruitment through the endothelium. Immunology. 2010;131(3):357–70.
136. Rainger GE, Nash GB. Cellular pathology of atherosclerosis: smooth muscle cells prime cocultured endothelial cells for enhanced leukocyte adhesion. Circ Res. 2001;88(6):615–22.
137. Filer A, Ward LSC, Kemble S, Davies CS, Munir H, Rogers R, et al. Identification of a transitional fibroblast function in very early rheumatoid arthritis. Ann Rheum Dis. 2017;76(12):2105–12.
138. McGettrick HM, Butler LM, Buckley CD, Ed Rainger G, Nash GB. Tissue stroma as a regulator of leukocyte recruitment in inflammation. J Leukoc Biol. 2012;91:385–400.
139. National Institutes of Health US. https://clinicaltrials.gov/
140. Weng JY, Du X, Geng SX, Peng YW, Wang Z, Lu ZS, et al. Mesenchymal stem cell as salvage treatment for refractory chronic GVHD. Bone Marrow Transplant. 2010;45(12):1732–40.
141. Ringden O, Uzunel M, Rasmusson I, Remberger M, Sundberg B, Lonnies H, et al. Mesenchymal stem cells for treatment of therapy-resistant graft-versus-host disease. Transplantation. 2006;81(10):1390–7.
142. Le Blanc K, Frassoni F, Ball L, Locatelli F, Roelofs H, Lewis I, et al. Mesenchymal stem cells for treatment of steroid-resistant, severe, acute graft-versus-host disease: a phase II study. Lancet. 2008;371(9624):1579–86.
143. Le Blanc K, Rasmusson I, Sundberg B, Gotherstrom C, Hassan M, Uzunel M, et al. Treatment of severe acute graft-versus-host disease with third party haploidentical mesenchymal stem cells. Lancet. 2004;363(9419):1439–41.
144. von Bonin M, Stolzel F, Goedecke A, Richter K, Wuschek N, Holig K, et al. Treatment of refractory acute GVHD with third-party MSC expanded in platelet lysate-containing medium. Bone Marrow Transplant. 2009;43(3):245–51.
145. Arima N, Nakamura F, Fukunaga A, Hirata H, Machida H, Kouno S, et al. Single intraarterial injection of mesenchymal stromal cells for treatment of steroid-refractory acute graft-versus-host disease: a pilot study. Cytotherapy. 2010;12:265–8.
146. Kurtzberg J, Prockop S, Teira P, Bittencourt H, Lewis V, Chan KW, et al. Allogeneic human mesenchymal stem cell therapy (remestemcel-L, Prochymal) as a rescue agent for

severe refractory acute graft-versus-host disease in pediatric patients. Biol Blood Marrow Transplant. 2014;20(2):229–35.

147. Bernardo ME, Ball LM, Cometa AM, Roelofs H, Zecca M, Avanzini MA, et al. Co-infusion of ex vivo-expanded, parental MSCs prevents life-threatening acute GVHD, but does not reduce the risk of graft failure in pediatric patients undergoing allogeneic umbilical cord blood transplantation. Bone Marrow Transplant. 2011;46(2):200–7.

148. Ning H, Yang F, Jiang M, Hu L, Feng K, Zhang J, et al. The correlation between cotransplantation of mesenchymal stem cells and higher recurrence rate in hematologic malignancy patients: outcome of a pilot clinical study. Leukemia. 2008;22(3):593–9.

149. Forslow U, Blennow O, LeBlanc K, Ringden O, Gustafsson B, Mattsson J, et al. Treatment with mesenchymal stromal cells is a risk factor for pneumonia-related death after allogeneic hematopoietic stem cell transplantation. Eur J Haematol. 2012;89(3):220–7.

150. Fu Y, Yan Y, Qi Y, Yang L, Li T, Zhang N, et al. Impact of autologous mesenchymal stem cell infusion on neuromyelitis optica spectrum disorder: a pilot, 2-year observational study. CNS Neurosci Ther. 2016;22(8):677–85.

151. Munir H, McGettrick HM. Mesenchymal stem cells therapy for autoimmune disease: risks and rewards. Stem Cells Dev. 2015;24(18):2091–100.

152. von Bahr L, Sundberg B, Lönnies L, Sander B, Karbach H, Hägglund H, et al. Long-term complications, immunologic effects, and role of passage for outcome in mesenchymal stromal cell therapy. Biol Blood Marrow Transplant. 2012;18:557–64.

153. Eggenhofer E, Benseler V, Kroemer A, Popp FC, Geissler EK, Schlitt HJ, et al. Mesenchymal stem cells are short-lived and do not migrate beyond the lungs after intravenous infusion. Front Immunol. 2012;3:297.

154. de Witte SFH, Luk F, Sierra Parraga JM, Gargesha M, Merino A, Korevaar SS, et al. Immunomodulation by therapeutic mesenchymal stromal cells (MSC) is triggered through phagocytosis of MSC by monocytic cells. Stem Cells. 2018;36(4):602–15.

155. Galleu A, Riffo-Vasquez Y, Trento C, Lomas C, Dolcetti L, Cheung TS, et al. Apoptosis in mesenchymal stromal cells induces in vivo recipient-mediated immunomodulation. Sci Transl Med. 2017;9(416):eaam7828.

Chapter 6
Stromal Cells in the Tumor Microenvironment

Alice E. Denton, Edward W. Roberts, and Douglas T. Fearon

Abstract The tumor microenvironment comprises a mass of heterogeneous cell types, including immune cells, endothelial cells, and fibroblasts, alongside cancer cells. It is increasingly becoming clear that the development of this support niche is critical to the continued uncontrolled growth of the cancer. The tumor microenvironment contributes to the maintenance of cancer stemness and also directly promotes angiogenesis, invasion, metastasis, and chronic inflammation. In this chapter, we describe on the role of fibroblasts, specifically termed cancer-associated fibroblasts (CAFs), in the promotion and maintenance of cancers. CAFs have a multitude of effects on the growth and maintenance of cancer, and here we focus on their roles in modulating immune cells and responses; CAFs both inhibit immune cell access to the tumor microenvironment and inhibit their functions within the tumor. Finally, we describe the potential modulation of CAF function as an adjunct to bolster the effectiveness of cancer immunotherapies.

Keywords Cancer-associated fibroblast · Tumor microenvironment · Stromal immunology · Immunotherapy

A. E. Denton (✉)
Lymphocyte Signalling and Development, Babraham Institute, Cambridge, UK
e-mail: alice.denton@babraham.ac.uk

E. W. Roberts
Department of Pathology, University of California San Francisco, San Francisco, CA, USA
e-mail: Edward.roberts@ucsf.edu

D. T. Fearon
Cold Spring Harbor Laboratory, Weill Cornell Medical College, New York, NY, USA
e-mail: dfearon@cshl.edu

© Springer International Publishing AG, part of Springer Nature 2018
B. M. J. Owens, M. A. Lakins (eds.), *Stromal Immunology*, Advances in Experimental Medicine and Biology 1060, https://doi.org/10.1007/978-3-319-78127-3_6

6.1 Introduction

The critical role of the tumor microenvironment in carcinogenesis has long been recognized, with Virchow first noting that malignancy arose at sites of chronic inflammation in 1863 [1]. In 1889 a more holistic hypothesis, that of the "seed and soil," was proposed by Paget suggesting that elements of the stroma were important for tumor development [2]. At a similar time, physicians noted that the status of the immune system may have important consequences for tumor development. Indeed sarcoma remissions had been observed following *Streptococcus pyogenes* infections, and in 1868 Busch induced a local infection and demonstrated a reduction in tumor burden, at least while the infection was ongoing [3]. Later several other physicians demonstrated remissions through localized applications of infectious agents [4–6]. However, these observations were soon overshadowed by other contemporary discoveries, namely, the identification of tumor suppressor genes and oncogenes as well as the advent of chemotherapy and radiotherapy. These led to increased focus on cell intrinsic mechanisms of carcinogenesis and more easily controlled therapeutic options, respectively, consigning the immune system and the stroma to the backburner.

6.2 Fibroblasts

There has, however, been a resurgence of interest in the tumor stroma in more recent times. In 1982 Bissell et al. outlined a modern formulation of Paget's seed and soil hypothesis stating that the tumor microenvironment is as important for tumor development as the accumulation of enabling genetic mutations [7]. This was based on the several elegant experiments showing that normalization of the stromal microenvironment could suppress tumor development; Illmensee and Mintz showed that while teratocarcinoma cells could be repeatedly transplanted and grow as ascites tumors, they would contribute to normal tissues when injected into a blastocyst [8]. Bissell's group also showed that normalization of integrin signaling in both 3D culture and in vivo could abrogate the malignant phenotype of genetically deranged breast cancer cell lines [9]. This focus on the extracellular matrix (ECM) and its role in modulating tumor cell behavior led to a general interest in cells responsible for generating and modulating the collagen matrix-fibroblasts. First described in 1858 based on their morphology and location [10], fibroblasts are non-epithelial, nonvascular, and non-hematopoietic cells in the connective tissue and are largely responsible for the synthesis of the ECM [11].

Fibroblasts are critical in both normal homeostasis and during wound healing. At steady state, fibroblasts are essential for epithelial homeostasis in many normal tissues having both direct interactions with the epithelial cells and secreting growth factors [12]. During wound healing macrophages produce transforming growth factor-β (TGF-β) and platelet-derived growth factor (PDGF) leading to activation of normal tissue fibroblasts [13, 14] to acquire a myofibroblast state defined by expression of α-smooth muscle actin (α-SMA) [15]. These myofibroblasts, first described in granulation tissue [16], play important roles in wound healing. Early histological studies showed that tumors appear similarly to healing wounds with Dvorak describing them as "wounds which do not heal" [17]; in the context of a healing wound angiogenesis, remodeling of the ECM and epithelial proliferation are all adaptive; however in the tumor microenvironment, these aid in tumor development. As may be expected in the tumor microenvironment, similar processes are ongoing as in healing wounds, and carcinoma-associated fibroblasts (CAFs) have also previously been defined by their expression of α-SMA [18]. Indeed the roles outlined above are all subverted during tumor progression to facilitate continued growth. In this context there has been great attention to the roles CAFs play compared to normal tissue fibroblasts; however, it has been challenging to study these cells in vivo due to the absence of ideal markers of these cells and the heterogeneity of the CAF population.

6.3 CAFs

Markers used to identify CAFS are often controversial. Fibroblast-specific protein-1 (FSP-1), one marker widely used to identify CAFs, has also been shown to be expressed by monocytes and invasive carcinoma cells [19, 20]; α-SMA on the other hand is expressed by pericytes and vascular smooth muscle cells [21, 22], while platelet-derived growth factor receptor-α (PDGFR-α) marks normal tissue fibroblasts [23] and some non-fibroblastic cells in the retinal pigment epithelium (unpublished observations and [24]). One promising marker for CAFs was fibroblast activation protein-α (FAP-α), which was identified by its reactivity with the F19 monoclonal antibody and reported to be selectively expressed on fibroblasts in healing wounds and in adenocarcinomas [25]. Further study appeared to support this with FAP-expressing cells being found in chronic inflammatory situations [26, 27]; however, when a reporter of FAP was generated, it was found to also be expressed on normal tissue fibroblasts, on fibroblastic reticular cells of the lymph node, and on some epithelial cells in the retinal pigment epithelium ([28, 29], unpublished observations) indicating that this too suffered from similar limitations regarding specificity. As such there has been a general lack of genetic systems by which to alter fibroblast function in vivo to dissect these roles more precisely. Furthermore, in many studies CAFs have been treated as a single entity, which is likely an oversimplification. Even normal tissue fibroblasts display remarkable heterogeneity with fibroblasts isolated from skin at different sites being as different transcriptomically as different leukocyte subsets [30]. Kidd et al. demonstrated in 2012 that fibroblast subsets defined by different markers originated from different sites with FSP-1+ fibroblasts deriving from the bone marrow, while other stromal cells may originate from adjacent tissues [31]. Indeed CAFs have many potential origins but most are thought to arise from local progenitors. CAFs are most commonly derived from tissue-resident fibroblasts [18, 32, 33], which are induced to undergo activation in response to the tumor microenvironment produced by neoplastic cells [34–38]. CAFs can also be derived from stellate cells [39, 40], migration of adipose or bone marrow-derived stromal cells [31, 41, 42], and endothelial and epithelial cells, through endothelial- or epithelial-to-mesenchymal transition [43, 44]. Despite these caveats there have been many successful ingenious studies elucidating the critical roles CAFs play in tumor development.

6.4 Driving Cancer Cell Proliferation

As stated earlier fibroblasts support epithelial homeostasis through secretion of growth factors, and CAFs too have been shown to have a direct effect on cancer cell growth in some systems. Orimo et al. showed that CXCL12 produced by CAFs drove cancer cell growth through CXCR4 expressed on the cancer cells [45]. CAFs have also been shown to produce numerous growth factors in different systems including insulin-like growth factors [46], connective tissue growth factor (CTGF) [47], platelet-derived growth factor (PDGF) [48], and hepatocyte growth factor [49], which have been shown to stimulate tumor cell growth in vitro. There is also evidence that CAFs can stimulate cell growth by releasing growth factors from the ECM through expression of matrix metalloproteases (MMPs). This indirect effect of MMP expression is a character shared with tumor-associated macrophages (TAMs) and other stromal cells [50]. CAFs do not simply affect the growth of cancer cells; however, they also modulate their phenotype making them more carcinogenic. Wang et al. showed that injection of an SV40-immortalized but not transformed prostate cancer cell line with CAFs led to poorly differentiated carcinomas developing, while there was minimal growth and no tumor development when this was carried out with normal prostate fibroblasts. When these cells were injected along with urogenital mesenchyme, epithelial cell growth was observed although there was no tumor development. As a result this series of experiments suggested that while the CAFs did stimulate growth, they had other pro-tumorigenic effects distinct from stimulating cell division [51].

6.5 Maintaining Cancer Stemness

Recent work has suggested that CAFs may also have a role in maintaining the "stemness" of cancer stem cells. Work in the intestinal crypt has shown that Wnt signaling is important in the maintenance of stem cells and crypt homeostasis [52]. Vermeulen et al. used a reporter of β-catenin-driven transcription to show that there was heterogenous Wnt signaling within colon cancer spheroidal cultures despite all cancer cells having mutations in the APC gene. The cells with higher levels of Wnt signaling were shown to be "cancer stem cells," that is, they had enhanced clonogenicity and were able to recreate a tumor similar to the initial malignancy if injected into an immune-deficient mouse. As there was heterogeneity in the spheroids, it was apparent that there was some cell autonomous regulation of Wnt signaling, and this has been shown previously in stem cells in the crypts [53]. However Wnt signaling in both the normal and the cancer stem cells is also modulated by surrounding cells [54]. Subsequently it was shown that HGF produced by CAFs induced high levels of Wnt signaling and increased "stemness" [55]. Thus in at least one system, CAFs have been reported to be important for the maintenance of cancer "stemness."

6.6 Driving Angiogenesis

In their study mentioned earlier, Orimo et al. isolated fibroblasts from human breast carcinomas and normal fibroblasts from the same patients. These fibroblasts were co-injected with breast carcinoma cells into nude mice, and it was shown that CAFs promoted tumor growth significantly more than did the normal fibroblasts. This was shown to be due, at least partially, to the high levels of CXCL12 secreted by CAFs which recruited endothelial progenitors, increasing vascularization of the tumors [45]. Furthermore Yang et al. subsequently showed that CAFs isolated from human prostate cancer similarly promoted xenograft growth due to the action of CTGF. It was found that CTGF expression was induced by TGFβ and that overexpression of CTGF in 3T3 fibroblasts also led to these cells increasing tumor size and microvessel density in a xenograft model [56]. CAFs can also promote angiogenesis in a less direct method by releasing active growth factors from the ECM due to their expression of MMPs. CAFs are a source of MMP9 [57] and MMP13 [58], which have both been shown to be involved in angiogenesis. MMP9 and MMP13 have both been shown to release vascular endothelial growth factors (VEGF) from the ECM increasing angiogenesis in the tumor [59, 60]. This work is complicated by the fact that in integrin α1 knock out mice which lack integrin α1β1, an inhibitor of MMP synthesis, there is decreased tumor vascularization due to the increased production of angiostatin [61]. Angiostatin is produced by MMP9 and MMP7 acting on circulating plasminogen. As such MMPs have been shown to have conflicting roles in angiogenesis. As stated previously MMPs are not restricted to CAFs as TAMs, and other stromal cells are also important sources of these molecules [62].

6.7 Promoting Invasion and Metastasis

CAFs may also exert their effects through modulation of ECM composition; Levental et al. showed that CAFs express lysyl oxidase, which leads to collagen cross-linking and increased tissue stiffness. This increased stiffness was associated with changes in integrin signaling and progression to invasive disease, and treatment of the mammary fat pad with lysyl oxidase could promote growth and invasion of premalignant cells highlighting the critical role for these stromal cells

[63]. CAFs have also been shown to express multiple MMPs that, by altering interactions between tumor cells and the extracellular matrix, alter tumor cell phenotype [57, 64, 65]. MMP activity has been implicated in all of the functions of CAFs so far, and it has been demonstrated that activation of MMPs is sufficient to produce a CAF-like phenotype in fibroblasts [66]. In this study, by deleting all four tissue inhibitors of metalloproteinases (TIMPs), the authors demonstrated that exosomes from the CAFs induced cancer cell motility and upregulated stem cell markers. These effects were dependent on the metalloproteinase ADAM10 [66]. Fibroblast exosomes have been shown to drive Wnt-planar cell polarity signaling in a CD81-dependent manner. This promoted breast cancer cell invasive behavior [67]. Interestingly communication between cancer cells and stromal components using exosomes appears to occur in both directions to promote metastasis with transfer of miR-105 from cancer cells to endothelial cells leading to increased vascular permeability and metastasis [68]. Furthermore cancer-derived exosomes appear to have roles in transmitting invasive behaviors between different cancer cell clones [69]. Intriguingly it appears that stromal cells may even play a more direct role in metastasis, with tumor cells which were part of heterotypic cell clusters demonstrating increased robustness and increased potential to develop metastases. Furthermore using a parabiosis model CAFs from the original tumor could transiently be found in metastases indicating these cells could be important in establishing metastases [70].

6.8 Promoting Resistance to Therapy

CAFs have also been implicated in driving resistance to chemotherapy. Previous work using the KPC mouse has shown that the chemotherapeutic drug gemcitabine is excluded due to dense ECM and high intratumoral tissue pressure. Enzymatically disrupting the ECM led to increased infiltration of the tumor and increased response [71, 72]. While this is a presumed effect of CAFs due to their role in producing the dense desmoplastic stroma in these tumors, fibroblast-derived exosomes have been shown to directly promote resistance to chemotherapy. One study demonstrated that exosomes carried numerous RNAs which stimulated RIG-I in breast cancer cells and along with NOTCH3 signaling driven by the CAFs themselves converged on STAT1 signaling which led to expansion of tumor cells resistant to both chemotherapy and radiotherapy [73].

6.9 CAFs and Inflammation

There have been also been unbiased approaches to understand how CAFs differ from normal fibroblasts in an attempt to define their roles in cancer. FAP+ cells sorted from normal tissues were found by RNA-seq to be similar to one another, while different CAF populations characterized by co-expression or lack of CD34 were found to be more similar to one another validating this as an approach [74]. Hanahan conducted a more complete analysis sorting PDGFR-α+ cells from a range of normal tissues and from tumors and demonstrated a CAF-specific gene signature [23]. This gene signature was characterized by CAFs producing greater amounts of CXCL1, CXCL2, IL-1β, and IL-6 (among others) than fibroblasts in normal tissues. This inflammatory gene signature implies that CAFs could be important players within the tumor inflammatory environment. Previous work has shown that inflammatory mediators play important roles in carcinogenesis: IL-6 may also protect tumor cells from apoptosis in a STAT3-dependent mechanism [75] and can drive angiogenesis [76], and IL-1β has also been shown to drive IL-23 expression and thus to promote skin carcinogenesis [77]. Despite these more direct effects, the inflammatory signature also suggests that CAFs may be modulating the immune system in the tumor microenvironment.

6.10 Tumor Immunology

While the role for the tumor stroma in tumor development was becoming more widely accepted, the role for the immune system in tumor development was controversial until more recently. Indeed in 2000 Hanahan and Weinberg wrote a highly influential review about the hallmarks of cancer noting that there was an overly narrow focus on the genetically deranged cancer cells and that heterotypic signaling with normal stromal cells explained at least some of the aforementioned hallmarks of cancer [78]; however, it wasn't until 2011 when they penned an update that evading the immune system appeared as a critical hallmark of tumor development [79]. Interest in the immune response to cancer reemerged as more focused approaches to immune modulation began to demonstrate results. After the initial demonstration that blockade of CTLA-4 could induce rejection of a transplanted primary tumor and lead to protection from a rechallenge [80], immune therapies gradually made their way to the clinic. Recently numerous immunotherapeutic approaches to tumor therapy have shown dramatic responses in patients resulting in numerous approvals of checkpoint blockade targeting CTLA-4 and PD-1 [81–87], as well as the use of chimeric antigen receptor (CAR) T cells to target tumor stroma [88]. It is now clear that vast complement of immune cells populate tumors, including dendritic cells, macrophages, neutrophils, natural killer cells, mast cells, B cells, CD8+ T cells, CD4+ T cells (and the many subsets thereof), and regulatory T cells (Treg cells). These cells have multiple roles in tumor and can have both pro- and antitumoral effects. These effects are, at least in part, directed by the microenvironment and are a product of CAFs and the inflammatory milieu they contribute to.

6.11   Stromal and Tumor Immune Responses

While these two areas have experienced renewed interest, interactions between these fields have only begun to emerge more recently. One of the most well-defined roles CAFs play in suppressing the antitumoral immune response regards entry of T cells. T cell infiltration into cancer nests within the tumor has long been recognized as an important predictor of patient survival, with increased infiltration of CD8+ T cells into cancer-dense regions strongly associated with improved outcomes for the patient [89–92]. Indeed, recent clinical trials using checkpoint blockade inhibitors and/or adoptive T cell therapy do not show strong results in tumors typically associated with high stromal burden, such as pancreatic ductal adenocarcinoma, prostate cancer, ovarian cancer, and colorectal cancer [82, 83, 93], in part due to a failure of CD8+ T cell to infiltrate cancer nests [94]. Normally, once activated, T cells leave the lymph node and migrate toward the inflammatory site, where they exit the blood stream and enter the tumor. The tumor vasculature inhibits extravasation of T cells from the blood stream into the tumor mass, resulting in accumulation of T cells within the stroma while still allowing movement of monocytes and neutrophils into the tumor. Indeed, overexpression of endothelin B receptor on tumor vasculature decreases lymphocyte adherence to endothelium [95], while the disorganization of tumor vasculature that is typically associated with tumor progression [96] limits T cell extravasation and entry into the tumor parenchyma [97].

6.12   CAFs and T Cell Infiltration

CAFs themselves may also directly limit T cell infiltration of cancer nests within the tumor. CAFs both deposit and degrade extracellular matrix (ECM) components and thus remodel the ECM during cancer progression; a severe desmoplastic reaction correlates with poor prognosis in many cancers [98–101]. The remodeling of the ECM allows CAFs to determine the migration and localization of cells within the tumor. Live cell imaging of human lung cancer [102] demonstrated poor T cell infiltration and motility in the dense collagen surrounding cancer nests, while more active T cell behavior was observed in regions with looser matrix deposition. Treatment with collagenase degraded the dense matrix surrounding cancer nests, and increased T cell infiltration and contact with cancer cells, suggesting that the nature of matrix deposition can have profound effects on the efficacy of antitumor T cell immunity. Recently, Fearon and colleagues demonstrated that, through production of CXCL12, pancreatic cancer-associated CAFs inhibit T cell infiltration of pancreatic carcinoma [74]. In this study, administration of checkpoint blockade inhibitors alone did not alter the course of tumor progression, mirroring that observed for human pancreatic cancer [93]. However, administration of the CXCR4 antagonist AMD3100 alongside checkpoint inhibitors significantly diminished tumor growth and allowed T cell infiltration and killing of cancer cells. Overexpression of

CXCL12 by CAFs [103] has been shown to promote cancer growth through direct effects on the cancer cells, promoting cancer cell proliferation, migration, and invasion [104–109]. However, since no change in tumor growth was observed in the absence of a T cell response, CXCL12 blockade must be affecting the immune regulation rather than the tumor-promoting aspect of CXCL12-CXCR4 signaling. Importantly, these studies have shown that the pre-existing T cell response is capable of inducing tumor regression when suppression is alleviated, suggesting that vaccination again tumor antigens will not be necessary to harness the antitumor immune response for immunotherapy.

6.13 Suppressing Intratumoral T Cells

Once T cells enter a tumor, they must migrate toward the cancer cells, engage the T cell receptor, and then deliver cytotoxic- and/or cytokine-mediated kill signals. There are several obstacles that T cells must overcome in the tumor microenvironment in order to achieve their aim. The tumor microenvironment is full of suppressive signals for T cells, including secreted factors, suppressive immune cells, and the immunosuppressive actions of CAFs. CAFs have been shown to exert a directly immunosuppressive mechanism of action. Depletion of FAP-expressing stromal cells using a diphtheria toxin-mediated model results in failure of tumor growth that is entirely dependent on the presence of an intact immune response, demonstrating that CAFs are a critical regulator of tumoral immunosuppression of the T cell response [110]. Tumor killing in this model was dependent on interferon-gamma and TNF-alpha, and was induced by hypoxic necrosis, all hallmarks of T cell-mediated immunity. Some factors produced by CAFs have also been identified; CAFs, along with myeloid cells and cancer cells, are important sources of tumoral indoleamine 2,3-dioxygenase (IDO), an enzyme that depletes local tryptophan by catabolizing tryptophan through the kynurenine pathway. Overexpression of IDO in the tumor microenvironment promotes tumor growth through immune resistance, as both T cell- and NK cell-mediated antitumoral responses are severely dysfunctional in the presence of IDO [111, 112]. CAFs also secrete TGFβ, and tumoral TGFβ expression is associated with significant pro-tumoral effect. TGFβ reduces CD8+ T cell and NK cell function [113–115], promotes the polarization of macrophages and neutrophils to a pro-tumoral type 2 phenotype [116–118], and enhances Treg cell and Th17 cell differentiation [119–121]. It is important to note that CAFs are not the sole source of TGFβ in the tumor microenvironment, with both neoplastic cells and myeloid-derived cells expressing high amounts of this cytokine. Accordingly, TGFβ blockade significantly inhibits tumor progression and enhances tumor immunotherapy [122, 123]. In addition to these effects, CAFs have been shown to promote the accumulation of immature myeloid cells, or MDSCs [124], which produce large amounts of suppressive cytokines, such as TGFβ, into the tumor microenvironment.

6.14　Skewing T Cell Recruitment

Alteration of CAFs by cancer cells has also been shown to alter the immune cell components within the tumor. For example, expression of the lymph node chemoattractant chemokine CCL21 in cancer cells, which is induced in invasive cancer cells [125], resulted stromal organization reminiscent of fibroblastic reticular cells (FRCs) of the lymph node. This FRC-like stromal organization recruited CCR7-expressing DCs, naïve T cells, and Treg cells to the tumor, leading to enhanced tumor growth through suppressive mechanisms including Treg cells and secreted molecules such as IDO [126]. Indeed, suppression was so profound it was able to prevent rejection of non-syngeneic allografts, demonstrating the immense capability of the tumor microenvironment to prevent T cell-mediated eradication of tumors. While this has not been shown to occur in human tumors, it points to the intriguing possibility that fibroblast populations as well as being heterogeneous are incredibly plastic, and our understanding of CAFs may be increased by investigating their roles in normal tissues like the lymph node.

6.15　Conclusions

There has been great interest in the potential of targeting CAFs in order to improve the response to immunotherapy in the clinic. While the responses to checkpoint blockade have been impressive with a proportion of patients showing complete, durable remissions [87], there are still many patients who do not respond. While direct ablation of CAFs led to tumor regression with immunogenic tumors [74, 110], the lack of specificity for the tumor led to significant side effects including cachexia and anemia when these cells were eliminated through administration of diphtheria toxin or by targeting with a FAP-specific chimeric antigen receptor T cell [29, 127]. As stated earlier there is a lack of markers that are specific for CAFs, and as such this type of approach is currently not feasible, and so greater understanding of the mechanisms by which CAFs exert their effects is needed in order to develop more targeted therapeutics. A promising example is the previously mentioned study by Feig et al. demonstrating that AMD3100 could potentiate the activity of checkpoint blockade [74]. As previously stated the heterogeneity of normal tissue fibroblasts [30] would suggest that while Hanahan found a CAF signature across different tumor types [23], there is still likely to be great heterogeneity in the mechanisms of immune suppression that CAFs are employing in different contexts. As such it is unlikely that there is any single CAF mechanism of action that can be targeted across all tumor types. It is critical that future work continues to elucidate the essential roles CAFs are playing in immune modulation in different tumor types, and it is also crucial that findings in mouse models are extended into patient samples to examine whether these potential therapeutic directions are viable.

References

1. Balkwill F, Mantovani A. Inflammation and cancer: back to Virchow? Lancet. 2001;357(9255):539–45.
2. Paget S. The distribution of secondary growths in cancer of the breast. 1889. Cancer Metastasis Rev. 1989;8(2):98–101.
3. Busch W. Aus der sitzung der medicinischen section vom 13. Berliner Klin Wochenschrift. 1868;5:137.
4. Fehleisen F. Ueber die Zuchtung der Erysipelkokken auf kunstlichem Nahrboden und die Ubertragbarkeit auf den Menschen. Dtsch Med Wochenschr. 1882;8:553–4.
5. Bruns P. Die Heilwirkung des Erysipelas auf Geschwuelste. Beitr Klin Chir. 1888;3:443
6. Coley WB. A preliminary note on the treatment of inoperable sarcoma by the toxic product of erysipelas. Postgrad Dent. 1893;8:278.
7. Bissell MJ, Hall HG, Parry G. How does the extracellular matrix direct gene expression? J Theor Biol. 1982;99(1):31–68.
8. Illmensee K, Mintz B. Totipotency and normal differentiation of single teratocarcinoma cells cloned by injection into blastocysts. Proc Natl Acad Sci U S A. 1976;73(2):549–53.
9. Weaver VM, Petersen OW, Wang F, Larabell CA, Briand P, Damsky C, et al. Reversion of the malignant phenotype of human breast cells in three-dimensional culture and in vivo by integrin blocking antibodies. J Cell Biol. 1997;137(1):231–45.
10. Virchow R. Die Cellularpathologie in Ihrer Begruendung auf Physiologische und Pathologische Gewebelehre. Hirschwald, A, Berlin. 1858.
11. Tarin D, Croft CB. Ultrastructural features of wound healing in mouse skin. J Anat. 1969;105(Pt 1):189–90.
12. Bellusci S, Grindley J, Emoto H, Itoh N, Hogan BL. Fibroblast growth factor 10 (FGF10) and branching morphogenesis in the embryonic mouse lung. Development. 1997;124(23):4867–78.
13. Wahl SM, McCartney-Francis N, Allen JB, Dougherty EB, Dougherty SF. Macrophage production of TGF-beta and regulation by TGF-beta. Ann N Y Acad Sci. 1990;593:188–96.
14. Bonner JC, Osornio-Vargas AR, Badgett A, Brody AR. Differential proliferation of rat lung fibroblasts induced by the platelet-derived growth factor-AA, -AB, and -BB isoforms secreted by rat alveolar macrophages. Am J Respir Cell Mol Biol. 1991;5(6):539–47.
15. Skalli O, Ropraz P, Trzeciak A, Benzonana G, Gillessen D, Gabbiani G. A monoclonal antibody against alpha-smooth muscle actin: a new probe for smooth muscle differentiation. J Cell Biol. 1986;103(6 Pt 2):2787–96.
16. Ryan GB, Cliff WJ, Gabbiani G, Irle C, Montandon D, Statkov PR, et al. Myofibroblasts in human granulation tissue. Hum Pathol. 1974;5(1):55–67.
17. Dvorak HF. Tumors: wounds that do not heal. Similarities between tumor stroma generation and wound healing. N Engl J Med. 1986;315(26):1650–9.
18. Ronnov-Jessen L, Petersen OW. Induction of alpha-smooth muscle actin by transforming growth factor-beta 1 in quiescent human breast gland fibroblasts. Implications for myofibroblast generation in breast neoplasia. Lab Investig. 1993;68(6):696–707.
19. Le Hir M, Hegyi I, Cueni-Loffing D, Loffing J, Kaissling B. Characterization of renal interstitial fibroblast-specific protein 1/S100A4-positive cells in healthy and inflamed rodent kidneys. Histochem Cell Biol. 2005;123(4–5):335–46.
20. Strutz F, Okada H, Lo CW, Danoff T, Carone RL, Tomaszewski JE, et al. Identification and characterization of a fibroblast marker: FSP1. J Cell Biol. 1995;130(2):393–405.
21. Tomasek JJ, Gabbiani G, Hinz B, Chaponnier C, Brown RA. Myofibroblasts and mechano-regulation of connective tissue remodelling. Nat Rev Mol Cell Biol. 2002;3(5):349–63.
22. Ronnov-Jessen L, Petersen OW, Bissell MJ. Cellular changes involved in conversion of normal to malignant breast: importance of the stromal reaction. Physiol Rev. 1996;76(1):69–125.
23. Erez N, Truitt M, Olson P, Arron ST, Hanahan D. Cancer-associated fibroblasts are activated in incipient neoplasia to orchestrate tumor-promoting inflammation in an NF-kappaB-dependent manner. Cancer Cell. 2010;17(2):135–47.

24. Vinores SA, Henderer JD, Mahlow J, Chiu C, Derevjanik NL, Larochelle W, et al. Isoforms of platelet-derived growth factor and its receptors in epiretinal membranes: immunolocalization to retinal pigmented epithelial cells. Exp Eye Res. 1995;60(6):607–19.
25. Garin-Chesa P, Old LJ, Rettig WJ. Cell surface glycoprotein of reactive stromal fibroblasts as a potential antibody target in human epithelial cancers. Proc Natl Acad Sci U S A. 1990;87(18):7235–9.
26. Bauer S, Jendro MC, Wadle A, Kleber S, Stenner F, Dinser R, et al. Fibroblast activation protein is expressed by rheumatoid myofibroblast-like synoviocytes. Arthritis Res Ther. 2006;8(6):R171.
27. Levy MT, McCaughan GW, Abbott CA, Park JE, Cunningham AM, Muller E, et al. Fibroblast activation protein: a cell surface dipeptidyl peptidase and gelatinase expressed by stellate cells at the tissue remodelling interface in human cirrhosis. Hepatology. 1999;29(6):1768–78.
28. Denton AE, Roberts EW, Linterman MA, Fearon DT. Fibroblastic reticular cells of the lymph node are required for retention of resting but not activated CD8+ T cells. Proc Natl Acad Sci U S A. 2014;111(33):12139–44.
29. Roberts EW, Deonarine A, Jones JO, Denton AE, Feig C, Lyons SK, et al. Depletion of stromal cells expressing fibroblast activation protein-alpha from skeletal muscle and bone marrow results in cachexia and anemia. J Exp Med. 2013;210(6):1137–51.
30. Chang HY, Chi JT, Dudoit S, Bondre C, van de Rijn M, Botstein D, et al. Diversity, topographic differentiation, and positional memory in human fibroblasts. Proc Natl Acad Sci U S A. 2002;99(20):12877–82.
31. Kidd S, Spaeth E, Watson K, Burks J, Lu H, Klopp A, et al. Origins of the tumor microenvironment: quantitative assessment of adipose-derived and bone marrow-derived stroma. PLoS One. 2012;7(2):e30563.
32. Kojima Y, Acar A, Eaton EN, Mellody KT, Scheel C, Ben-Porath I, et al. Autocrine TGF-beta and stromal cell-derived factor-1 (SDF-1) signaling drives the evolution of tumor-promoting mammary stromal myofibroblasts. Proc Natl Acad Sci U S A. 2010;107(46):20009–14.
33. Quante M, Tu SP, Tomita H, Gonda T, Wang SS, Takashi S, et al. Bone marrow-derived myofibroblasts contribute to the mesenchymal stem cell niche and promote tumor growth. Cancer Cell. 2011;19(2):257–72.
34. Giannoni E, Bianchini F, Masieri L, Serni S, Torre E, Calorini L, et al. Reciprocal activation of prostate cancer cells and cancer-associated fibroblasts stimulates epithelial-mesenchymal transition and cancer stemness. Cancer Res. 2010;70(17):6945–56.
35. Lohr M, Schmidt C, Ringel J, Kluth M, Muller P, Nizze H, et al. Transforming growth factor-beta1 induces desmoplasia in an experimental model of human pancreatic carcinoma. Cancer Res. 2001;61(2):550–5.
36. Shao ZM, Nguyen M, Barsky SH. Human breast carcinoma desmoplasia is PDGF initiated. Oncogene. 2000;19(38):4337–45.
37. Yang G, Rosen DG, Zhang Z, Bast RC Jr, Mills GB, Colacino JA, et al. The chemokine growth-regulated oncogene 1 (Gro-1) links RAS signaling to the senescence of stromal fibroblasts and ovarian tumorigenesis. Proc Natl Acad Sci U S A. 2006;103(44):16472–7.
38. Lieubeau B, Garrigue L, Barbieux I, Meflah K, Gregoire M. The role of transforming growth factor beta 1 in the fibroblastic reaction associated with rat colorectal tumor development. Cancer Res. 1994;54(24):6526–32.
39. Bachem MG, Schneider E, Gross H, Weidenbach H, Schmid RM, Menke A, et al. Identification, culture, and characterization of pancreatic stellate cells in rats and humans. Gastroenterology. 1998;115(2):421–32.
40. Yin C, Evason KJ, Asahina K, Stainier DY. Hepatic stellate cells in liver development, regeneration, and cancer. J Clin Invest. 2013;123(5):1902–10.
41. Ishii G, Sangai T, Oda T, Aoyagi Y, Hasebe T, Kanomata N, et al. Bone-marrow-derived myofibroblasts contribute to the cancer-induced stromal reaction. Biochem Biophys Res Commun. 2003;309(1):232–40.
42. Direkze NC, Hodivala-Dilke K, Jeffery R, Hunt T, Poulsom R, Oukrif D, et al. Bone marrow contribution to tumor-associated myofibroblasts and fibroblasts. Cancer Res. 2004;64(23):8492–5.

43. Iwano M, Plieth D, Danoff TM, Xue C, Okada H, Neilson EG. Evidence that fibroblasts derive from epithelium during tissue fibrosis. J Clin Invest. 2002;110(3):341–50.
44. Zeisberg EM, Potenta S, Xie L, Zeisberg M, Kalluri R. Discovery of endothelial to mesenchymal transition as a source for carcinoma-associated fibroblasts. Cancer Res. 2007;67(21):10123–8.
45. Orimo A, Gupta PB, Sgroi DC, Arenzana-Seisdedos F, Delaunay T, Naeem R, et al. Stromal fibroblasts present in invasive human breast carcinomas promote tumor growth and angiogenesis through elevated SDF-1/CXCL12 secretion. Cell. 2005;121(3):335–48.
46. Yee D, Rosen N, Favoni RE, Cullen KJ. The insulin-like growth factors, their receptors, and their binding proteins in human breast cancer. Cancer Treat Res. 1991;53:93–106.
47. Frazier KS, Grotendorst GR. Expression of connective tissue growth factor mRNA in the fibrous stroma of mammary tumors. Int J Biochem Cell Biol. 1997;29(1):153–61.
48. Ponten F, Ren Z, Nister M, Westermark B, Ponten J. Epithelial-stromal interactions in basal cell cancer: the PDGF system. J Invest Dermatol. 1994;102(3):304–9.
49. Nakamura T, Matsumoto K, Kiritoshi A, Tano Y, Nakamura T. Induction of hepatocyte growth factor in fibroblasts by tumor-derived factors affects invasive growth of tumor cells: in vitro analysis of tumor-stromal interactions. Cancer Res. 1997;57(15):3305–13.
50. Kessenbrock K, Plaks V, Werb Z. Matrix metalloproteinases: regulators of the tumor microenvironment. Cell. 2010;141(1):52–67.
51. Wang Y, Sudilovsky D, Zhang B, Haughney PC, Rosen MA, Wu DS, et al. A human prostatic epithelial model of hormonal carcinogenesis. Cancer Res. 2001;61(16):6064–72.
52. Korinek V, Barker N, Moerer P, van Donselaar E, Huls G, Peters PJ, et al. Depletion of epithelial stem-cell compartments in the small intestine of mice lacking Tcf-4. Nat Genet. 1998;19(4):379–83.
53. Sato T, Vries RG, Snippert HJ, van de Wetering M, Barker N, Stange DE, et al. Single Lgr5 stem cells build crypt-villus structures in vitro without a mesenchymal niche. Nature. 2009;459(7244):262–5.
54. Yen TH, Wright NA. The gastrointestinal tract stem cell niche. Stem Cell Rev. 2006;2(3):203–12.
55. Vermeulen L, De Sousa EMF, van der Heijden M, Cameron K, de Jong JH, Borovski T, et al. Wnt activity defines colon cancer stem cells and is regulated by the microenvironment. Nat Cell Biol. 2010;12(5):468–76.
56. Yang F, Tuxhorn JA, Ressler SJ, McAlhany SJ, Dang TD, Rowley DR. Stromal expression of connective tissue growth factor promotes angiogenesis and prostate cancer tumorigenesis. Cancer Res. 2005;65(19):8887–95.
57. Boire A, Covic L, Agarwal A, Jacques S, Sherifi S, Kuliopulos A. PAR1 is a matrix metalloprotease-1 receptor that promotes invasion and tumorigenesis of breast cancer cells. Cell. 2005;120(3):303–13.
58. Vosseler S, Lederle W, Airola K, Obermueller E, Fusenig NE, Mueller MM. Distinct progression-associated expression of tumor and stromal MMPs in HaCaT skin SCCs correlates with onset of invasion. Int J Cancer. 2009;125(10):2296–306.
59. Bergers G, Brekken R, McMahon G, Vu TH, Itoh T, Tamaki K, et al. Matrix metalloproteinase-9 triggers the angiogenic switch during carcinogenesis. Nat Cell Biol. 2000;2(10):737–44.
60. Lederle W, Hartenstein B, Meides A, Kunzelmann H, Werb Z, Angel P, et al. MMP13 as a stromal mediator in controlling persistent angiogenesis in skin carcinoma. Carcinogenesis. 2010;31(7):1175–84.
61. Pozzi A, Moberg PE, Miles LA, Wagner S, Soloway P, Gardner HA. Elevated matrix metalloprotease and angiostatin levels in integrin alpha 1 knockout mice cause reduced tumor vascularization. Proc Natl Acad Sci U S A. 2000;97(5):2202–7.
62. Mantovani A, Schioppa T, Porta C, Allavena P, Sica A. Role of tumor-associated macrophages in tumor progression and invasion. Cancer Metastasis Rev. 2006;25(3):315–22.
63. Levental KR, Yu H, Kass L, Lakins JN, Egeblad M, Erler JT, et al. Matrix crosslinking forces tumor progression by enhancing integrin signaling. Cell. 2009;139(5):891–906.
64. Kalluri R, Zeisberg M. Fibroblasts in cancer. Nat Rev Cancer. 2006;6(5):392–401.
65. Lochter A, Galosy S, Muschler J, Freedman N, Werb Z, Bissell MJ. Matrix metalloproteinase stromelysin-1 triggers a cascade of molecular alterations that leads to stable epithelial-to-

mesenchymal conversion and a premalignant phenotype in mammary epithelial cells. J Cell Biol. 1997;139(7):1861–72.

66. Shimoda M, Principe S, Jackson HW, Luga V, Fang H, Molyneux SD, et al. Loss of the Timp gene family is sufficient for the acquisition of the CAF-like cell state. Nat Cell Biol. 2014;16

67. Luga V, Zhang L, Viloria-Petit AM, Ogunjimi AA., Inanlou MR, Chiu E, et al. Exosomes mediate stromal mobilization of autocrine Wnt-PCP signaling in breast cancer cell migration. Cell. Elsevier Inc. 2012;151(7):1542–56.

68. Zhou W, Fong MY, Min Y, Somlo G, Liu L, Palomares MR, et al. Cancer-secreted miR-105 destroys vascular endothelial barriers to promote metastasis. Cancer Cell. Elsevier Inc. 2014;25(4):501–15.

69. Zomer A, Maynard C, Verweij FJ, Kamermans A, Schäfer R, Beerling E, et al. In vivo imaging reveals extracellular vesicle-mediated phenocopying of metastatic behavior. Cell. 2015;161(5):1046–57.

70. Duda DG, Duyverman AM, Kohno M, Snuderl M, Steller EJ, Fukumura D, et al. Malignant cells facilitate lung metastasis by bringing their own soil. Proc Natl Acad Sci U S A. 2010;107(50):21677–82.

71. Jacobetz MA, Chan DS, Neesse A, Bapiro TE, Cook N, Frese KK, et al. Hyaluronan impairs vascular function and drug delivery in a mouse model of pancreatic cancer. Gut. 2013;62(1):112–20.

72. Provenzano PP, Cuevas C, Chang AE, Goel VK, Von Hoff DD, Hingorani SR. Enzymatic targeting of the stroma ablates physical barriers to treatment of pancreatic ductal adenocarcinoma. Cancer Cell. Elsevier Inc. 2012;21(3):418–29.

73. Boelens MC, Wu TJ, Nabet BY, Xu B, Qiu Y, Yoon T, et al. Exosome transfer from stromal to breast cancer cells regulates therapy resistance pathways. Cell. Elsevier Inc. 2014;159(3):499–513.

74. Feig C, Jones JO, Kraman M, Wells RJ, Deonarine A, Chan DS, et al. Targeting CXCL12 from FAP-expressing carcinoma-associated fibroblasts synergizes with anti-PD-L1 immunotherapy in pancreatic cancer. Proc Natl Acad Sci U S A. 2013;110(50):20212–7.

75. Naugler WE, Karin M. The wolf in sheep's clothing: the role of interleukin-6 in immunity, inflammation and cancer. Trends Mol Med. 2008;14(3):109–19.

76. Nilsson MB, Langley RR, Fidler IJ. Interleukin-6, secreted by human ovarian carcinoma cells, is a potent proangiogenic cytokine. Cancer Res. 2005;65(23):10794–800.

77. Langowski JL, Zhang X, Wu L, Mattson JD, Chen T, Smith K, et al. IL-23 promotes tumour incidence and growth. Nature. 2006;442(7101):461–5.

78. Hanahan D, Weinberg RA. The hallmarks of cancer. Cell. 2000;100(1):57–70.

79. Hanahan D, Weinberg RA. Hallmarks of cancer: the next generation. Cell. 2011;144(5):646–74.

80. Leach DR, Krummel MF, Allison JP. Enhancement of antitumor immunity by CTLA-4 blockade. Science (80-). 1996;271(5256):1734–6.

81. Hodi FS, O'Day SJ, McDermott DF, Weber RW, Sosman JA, Haanen JB, et al. Improved survival with ipilimumab in patients with metastatic melanoma. N Engl J Med. 2010;363(8):711–23.

82. Brahmer JR, Tykodi SS, Chow LQ, Hwu WJ, Topalian SL, Hwu P, et al. Safety and activity of anti-PD-L1 antibody in patients with advanced cancer. N Engl J Med. 2012;366(26):2455–65.

83. Topalian SL, Hodi FS, Brahmer JR, Gettinger SN, Smith DC, McDermott DF, et al. Safety, activity, and immune correlates of anti-PD-1 antibody in cancer. N Engl J Med. 2012;366(26):2443–54.

84. Powles T, Eder JP, Fine GD, Braiteh FS, Loriot Y, Cruz C, et al. MPDL3280A (anti-PD-L1) treatment leads to clinical activity in metastatic bladder cancer. Nature. 2014;515(7528):558–62.

85. Herbst RS, Soria JC, Kowanetz M, Fine GD, Hamid O, Gordon MS, et al. Predictive correlates of response to the anti-PD-L1 antibody MPDL3280A in cancer patients. Nature. 2014;515(7528):563–7.

86. Tumeh PC, Harview CL, Yearley JH, Shintaku IP, Taylor EJ, Robert L, et al. PD-1 blockade induces responses by inhibiting adaptive immune resistance. Nature. 2014;515(7528):568–71.

87. Sharma P, Allison JP. The future of immune checkpoint therapy. Science (80-). 2015;348(6230):56–61.
88. Rosenberg SA, Restifo NP. Adoptive cell transfer as personalized immunotherapy for human cancer. Science (80-). 2015;348(6230):62–8.
89. Naito Y, Saito K, Shiiba K, Ohuchi A, Saigenji K, Nagura H, et al. CD8+ T cells infiltrated within cancer cell nests as a prognostic factor in human colorectal cancer. Cancer Res. 1998;58(16):3491–4.
90. Zhang L, Conejo-Garcia JR, Katsaros D, Gimotty PA, Massobrio M, Regnani G, et al. Intratumoral T cells, recurrence, and survival in epithelial ovarian cancer. N Engl J Med. 2003;348(3):203–13.
91. Sato E, Olson SH, Ahn J, Bundy B, Nishikawa H, Qian F, et al. Intraepithelial CD8+ tumor-infiltrating lymphocytes and a high CD8+/regulatory T cell ratio are associated with favorable prognosis in ovarian cancer. Proc Natl Acad Sci U S A. 2005;102(51):18538–43.
92. Galon J, Costes A, Sanchez-Cabo F, Kirilovsky A, Mlecnik B, Lagorce-Pages C, et al. Type, density, and location of immune cells within human colorectal tumors predict clinical outcome. Science (80-). 2006;313(5795):1960–4.
93. Royal RE, Levy C, Turner K, Mathur A, Hughes M, Kammula US, et al. Phase 2 trial of single agent Ipilimumab (anti-CTLA-4) for locally advanced or metastatic pancreatic adenocarcinoma. J Immunother. 2010;33(8):828–33.
94. Breart B, Lemaitre F, Celli S, Bousso P. Two-photon imaging of intratumoral CD8+ T cell cytotoxic activity during adoptive T cell therapy in mice. J Clin Invest. 2008;118(4):1390–7.
95. Buckanovich RJ, Facciabene A, Kim S, Benencia F, Sasaroli D, Balint K, et al. Endothelin B receptor mediates the endothelial barrier to T cell homing to tumors and disables immune therapy. Nat Med. 2008;14(1):28–36.
96. Ryschich E, Schmidt J, Hammerling GJ, Klar E, Ganss R. Transformation of the microvascular system during multistage tumorigenesis. Int J Cancer. 2002;97(6):719–25.
97. Hamzah J, Jugold M, Kiessling F, Rigby P, Manzur M, Marti HH, et al. Vascular normalization in Rgs5-deficient tumours promotes immune destruction. Nature. 2008;453(7193):410–4.
98. Cardone A, Tolino A, Zarcone R, Borruto Caracciolo G, Tartaglia E. Prognostic value of desmoplastic reaction and lymphocytic infiltration in the management of breast cancer. Panminerva Med. 1997;39(3):174–7.
99. Maeshima AM, Niki T, Maeshima A, Yamada T, Kondo H, Matsuno Y. Modified scar grade: a prognostic indicator in small peripheral lung adenocarcinoma. Cancer. 2002;95(12):2546–54.
100. Tsujino T, Seshimo I, Yamamoto H, Ngan CY, Ezumi K, Takemasa I, et al. Stromal myofibroblasts predict disease recurrence for colorectal cancer. Clin Cancer Res. 2007;13(7):2082–90.
101. Kawase A, Ishii G, Nagai K, Ito T, Nagano T, Murata Y, et al. Podoplanin expression by cancer associated fibroblasts predicts poor prognosis of lung adenocarcinoma. Int J Cancer. 2008;123(5):1053–9.
102. Salmon H, Franciszkiewicz K, Damotte D, Dieu-Nosjean MC, Validire P, Trautmann A, et al. Matrix architecture defines the preferential localization and migration of T cells into the stroma of human lung tumors. J Clin Invest. 2012;122(3):899–910.
103. Allinen M, Beroukhim R, Cai L, Brennan C, Lahti-Domenici J, Huang H, et al. Molecular characterization of the tumor microenvironment in breast cancer. Cancer Cell. 2004;6(1):17–32.
104. Domanska UM, Timmer-Bosscha H, Nagengast WB, Oude Munnink TH, Kruizinga RC, Ananias HJ, et al. CXCR4 inhibition with AMD3100 sensitizes prostate cancer to docetaxel chemotherapy. Neoplasia. 2012;14(8):709–18.
105. Li X, Ma Q, Xu Q, Liu H, Lei J, Duan W, et al. SDF-1/CXCR4 signaling induces pancreatic cancer cell invasion and epithelial-mesenchymal transition in vitro through non-canonical activation of Hedgehog pathway. Cancer Lett. 2012;322(2):169–76.
106. Wang Z, Ma Q, Liu Q, Yu H, Zhao L, Shen S, et al. Blockade of SDF-1/CXCR4 signalling inhibits pancreatic cancer progression in vitro via inactivation of canonical Wnt pathway. Br J Cancer. 2008;99(10):1695–703.
107. Hall JM, Korach KS. Stromal cell-derived factor 1, a novel target of estrogen receptor action, mediates the mitogenic effects of estradiol in ovarian and breast cancer cells. Mol Endocrinol. 2003;17(5):792–803.

108. Muller A, Homey B, Soto H, Ge N, Catron D, Buchanan ME, et al. Involvement of chemokine receptors in breast cancer metastasis. Nature. 2001;410(6824):50–6.
109. Scotton CJ, Wilson JL, Scott K, Stamp G, Wilbanks GD, Fricker S, et al. Multiple actions of the chemokine CXCL12 on epithelial tumor cells in human ovarian cancer. Cancer Res. 2002;62(20):5930–8.
110. Kraman M, Bambrough PJ, Arnold JN, Roberts EW, Magiera L, Jones JO, et al. Suppression of antitumor immunity by stromal cells expressing fibroblast activation protein-alpha. Science (80-). 2010;330(6005):827–30.
111. Li T, Yang Y, Hua X, Wang G, Liu W, Jia C, et al. Hepatocellular carcinoma-associated fibroblasts trigger NK cell dysfunction via PGE2 and IDO. Cancer Lett. 2012;318(2):154–61.
112. Uyttenhove C, Pilotte L, Theate I, Stroobant V, Colau D, Parmentier N, et al. Evidence for a tumoral immune resistance mechanism based on tryptophan degradation by indoleamine 2,3-dioxygenase. Nat Med. 2003;9(10):1269–74.
113. Thomas DA, Massague J. TGF-beta directly targets cytotoxic T cell functions during tumor evasion of immune surveillance. Cancer Cell. 2005;8(5):369–80.
114. Gorelik L, Flavell RA. Immune-mediated eradication of tumors through the blockade of transforming growth factor-beta signaling in T cells. Nat Med. 2001;7(10):1118–22.
115. Lee JC, Lee KM, Kim DW, Heo DS. Elevated TGF-beta1 secretion and down-modulation of NKG2D underlies impaired NK cytotoxicity in cancer patients. J Immunol. 2004;172(12):7335–40.
116. Gidekel Friedlander SY, Chu GC, Snyder EL, Girnius N, Dibelius G, Crowley D, et al. Context-dependent transformation of adult pancreatic cells by oncogenic K-Ras. Cancer Cell. 2009;16(5):379–89.
117. Tsunawaki S, Sporn M, Ding A, Nathan C. Deactivation of macrophages by transforming growth factor-beta. Nature. 1988;334(6179):260–2.
118. Allavena P, Sica A, Garlanda C, Mantovani A. The Yin-Yang of tumor-associated macrophages in neoplastic progression and immune surveillance. Immunol Rev. 2008;222:155–61.
119. Trapani JA. The dual adverse effects of TGF-beta secretion on tumor progression. Cancer Cell. 2005;8(5):349–50.
120. Ghiringhelli F, Puig PE, Roux S, Parcellier A, Schmitt E, Solary E, et al. Tumor cells convert immature myeloid dendritic cells into TGF-beta-secreting cells inducing CD4+CD25+ regulatory T cell proliferation. J Exp Med. 2005;202(7):919–29.
121. Mangan PR, Harrington LE, O'Quinn DB, Helms WS, Bullard DC, Elson CO, et al. Transforming growth factor-beta induces development of the T(H)17 lineage. Nature. 2006;441(7090):231–4.
122. Kim S, Buchlis G, Fridlender ZG, Sun J, Kapoor V, Cheng G, et al. Systemic blockade of transforming growth factor-beta signaling augments the efficacy of immunogene therapy. Cancer Res. 2008;68(24):10247–56.
123. Terabe M, Ambrosino E, Takaku S, O'Konek JJ, Venzon D, Lonning S, et al. Synergistic enhancement of CD8+ T cell-mediated tumor vaccine efficacy by an anti-transforming growth factor-beta monoclonal antibody. Clin Cancer Res. 2009;15(21):6560–9.
124. Mace TA, Ameen Z, Collins A, Wojcik S, Mair M, Young GS, et al. Pancreatic cancer-associated stellate cells promote differentiation of myeloid-derived suppressor cells in a STAT3-dependent manner. Cancer Res. 2013;73(10):3007–18.
125. Shields JD, Fleury ME, Yong C, Tomei AA, Randolph GJ, Swartz MA. Autologous chemotaxis as a mechanism of tumor cell homing to lymphatics via interstitial flow and autocrine CCR7 signaling. Cancer Cell. 2007;11(6):526–38.
126. Shields JD, Kourtis IC, Tomei AA, Roberts JM, Swartz MA. Induction of lymphoidlike stroma and immune escape by tumors that express the chemokine CCL21. Science (80-). 2010;328(5979):749–52.
127. Tran E, Chinnasamy D, Yu Z, Morgan RA, Lee CC, Restifo NP, et al. Immune targeting of fibroblast activation protein triggers recognition of multipotent bone marrow stromal cells and cachexia. J Exp Med. 2013;210(6):1125–35.

Chapter 7
Immunosuppression by Intestinal Stromal Cells

Iryna V. Pinchuk and Don W. Powell

Abstract This chapter summarizes evidence that intestinal myofibroblasts, also called intestinal stromal cells, are derived in the adult from tissue mesenchymal stem cells under homeostasis and may be replenished by bone marrow mesenchymal stromal (stem) cells that are recruited after severe intestinal injury. A comparison of mechanism of immunosuppression or tolerance by adult intestinal stromal cells (myofibroblasts) is almost identical with those reported for mesenchymal stem cells of bone marrow origin. The list of suppression mechanisms includes PD-L1 and PD-L2/PD-1 immune checkpoint pathways, soluble mediator secretion, toll-like receptor-mediated tolerance, and augmentation of Treg cells. Further, both mesenchymal stem cells and intestinal stromal cells express an almost identical repertoire of CD molecules. Lastly, others have reported that isolate intestinal stromal cells are capable of differentiating into bone and less well into chondrocyte, but not into adipocytes, a finding that we have confirmed. These findings suggest that intestinal stromal cells (myofibroblasts) are partially differentiated adult, tissue-resident stem cells which are capable of exerting immune tolerance in the intestine. Their role in repair of inflammatory bowel disease and immune suppression in colorectal cancer needs further investigation.

Keywords Mesenchymal stem cells · Tissue-resident adult mesenchymal stem cells · Myofibroblasts · Immune tolerance · PD-L1 · PD-L2 · Toll-like receptors · Inflammatory bowel disease · Colorectal cancer

This research was supported by NIH – NCATS U54 TR001342-01 and NIDDK R01 DK103150.

I. V. Pinchuk
Departments of Internal Medicine, University of Texas Medical Branch, Galveston, TX, USA

Microbiology and Immunology, University of Texas Medical Branch, Galveston, TX, USA
e-mail: ivpinchu@utmb.edu

D. W. Powell (⊠)
Departments of Internal Medicine, University of Texas Medical Branch, Galveston, TX, USA

Neuroscience, Cell Biology and Anatomy, University of Texas Medical Branch, Galveston, TX, USA
e-mail: dpowell@utmb.edu

© Springer International Publishing AG, part of Springer Nature 2018
B. M.J Owens, M. A. Lakins (eds.), *Stromal Immunology*, Advances in Experimental Medicine and Biology 1060, https://doi.org/10.1007/978-3-319-78127-3_7

7.1 Introduction

There are a multitude of pleotropic functions of intestinal stromal cells (myofibro-blasts, fibroblasts, and pericytes) that have been discovered and investigated over the past 20 years. Previous reviews document knowledge about these cells up until the current era [1–7]. Our chapter defines the role of mucosal stromal cells in gut tolerogenic responses, including immunosuppression by B7 suppressor molecules (PD-L1 and PD-L2) which are present on these MHC class II-expressing antigen-presenting cells, immunosuppression by soluble mediators secreted by stromal cells, role in altering Th17 cell and Treg formation, and toll-like receptor-mediated modulation of immunosuppression. The PD-L1/PD-1 signaling pathways have recently become famous with the discovery of immune checkpoint therapy for can-cer, revolutionizing the field of oncologic immune therapy and bringing effective therapies to previously untreatable cancers [8–10]. Our interest in these molecules developed when we discovered that mucosal CD90$^+$ stromal cells in gastric, small intestinal, and colonic mucosa were novel, innate immune cells expressing MHC class II [11–13]. Seeking an antigen-presenting function for these cells, we found that the positive B7 co-stimulatory molecules CD80 and CD86 were not normally expressed, although CD86 could be demonstrated after engagement of T cells [13]. Nevertheless, the negative co-stimulatory molecules PD-L1 and PD-L2 were robustly constitutively expressed [14], suggesting that CD90$^+$ stromal cells were far more important in tolerance than in activation of immunity.

In attempting to understand why stromal cells of all types – intestinal, chondro-cytes, synovial, lung, and skin fibroblasts – might have such potent immunosuppres-sive functions [15], an attractive hypothesis was found in the emerging concepts of the origin of intestinal stromal cells and the idea that they might be derived from adult or tissue-resident, adult mesenchymal stem cells (tMSC) or from the recruit-ment of bone marrow-derived mesenchymal stromal cells (BMMSC). This was especially true since the mechanisms of MSC-mediated immunosuppression and the ability of MSC, like intestinal stromal cells, to switch from inflammation to sup-pression have been more recently become better understood [16–18]. Therefore, before describing the information we have learned about the immune suppressive role of intestinal stromal cells, we will briefly review the current understanding of the origin of intestinal stromal cells and of the immune functions of MSC.

7.2 Origin of Intestinal Stromal Cells

It was once thought that parenchymal mesodermal cells in the embryo originated from the neural crest [19]. However, more recent lineage tracing experiments have defined the mesothelium as the embryological origin of tissue parenchymal (myo) fibroblasts, perivascular pericytes, and vascular smooth muscle cells [20, 21]. In the adult, the discussion has centered on whether the origin of subepithelial stromal cells (myofibroblasts), during homeostasis or after tissue damage, is from a tissue-resident, adult mesenchymal stem cells (tMSC) or from engraftment of circulating bone marrow mesenchymal stem cell (BMMSC) [22]. Stappenbeck's laboratory presented evidence for a tMSC, identified by its avid expression of COX-2 (and thus prostaglandin secretion), located in the upper aspects of the lamina propria but seemingly homing to a pericryptal location adjacent to epithelial stem cells in response to toll-like receptor (TLR) signaling [23]. Prostaglandin secretion from these relocating cells was critical for repair of dextran sodium sulfate (DSS)-induced colitis and for repair of experimental colonic perforating wounds [23]. While these cells had the repertoire of stems cells, their origin remained unclear [22]. Strong evidence that these cells may be tMSC has recently been published by Worthely et al. [24] in impressive lineage-tracking experiments using Gremlin 1 (Grem 1) as a marker of subepithelial mesenchymal cells. Tamoxifen-induced expression in a Grem1-creER^T mouse identified single subepithelial cells in the small intestine isthmus, the region that serves as the transition from villi to crypts. These cells divided exceedingly slowly, incorporating BrdU over the course of a month and taking 3 months for these labeled cells to populate the entire pericryptal mesenchymal sheath and a year to completely populate the entire villus with smooth muscle α-actin-positive myofibroblasts and smooth muscle α-actin-negative, but not NG2-positive, stromal pericyte-like cells. These marked cells persisted for 2 years. Worthely named these cells intestinal reticular stem cells (iRSC) denoting the reticular network that they formed. This network was entirely distinct from the closely approximated s100b-/NES-positive glial network. Thus, this publication gives strong evidence for a slow cycling, tissue stem cell providing homeostasis for small intestinal epithelial and lamina propria small vessel function and for tissue structure. Although results of studies in colonic and gastric tissues were not reported, Worthely has stated that similar observations were made for colonic and gastric mucosa (personal communication).

While it is possible that under conditions of significant intestinal damage these tMSC might be called to the damaged area by chemotaxis to repopulate the myofibroblast/fibroblasts and pericytes network, an alternative mechanism for rapid repair is also possible: homing of BMMSC. As demonstrated first by Britten and colleagues [25], and reviewed in detail by Mifflin et al. [5], using the Y chromosome from male bone marrow infused into female recipients under conditions of significant tissue wounding, BMMSC may reconstitute 40–60% of subepithelial myofibroblasts and pericytes within 10 weeks of transplantation. A similar phenomenon has been shown for the cancer microenvironment where 20% of cancer-associated fibroblasts in colorectal cancers are derived from BMMSC [26, 27]. Thus, one might reasonably postulate that tMSC are responsible for the homeostasis of the intestinal epithelium and lamina propria architecture, but that BMMSC replenishment serves as the mechanism for more rapid repair after acute damaging disease or trauma (Fig. 7.1).

More compelling proof that intestinal stromal cells are derived from MSC comes from the study of Signore et al. [28] who used CD146, a known MSC marker, to visualize lamina propria cells by confocal microscopy and to isolate them. The CD146-positive cells have the location and appearance of colonic CD90$^+$ myofibroblasts/fibroblasts. Importantly, isolated colonic CD146 cells had the same marker phenotype as BMMSC but had a decreased intensity of the CD13, CD29, and

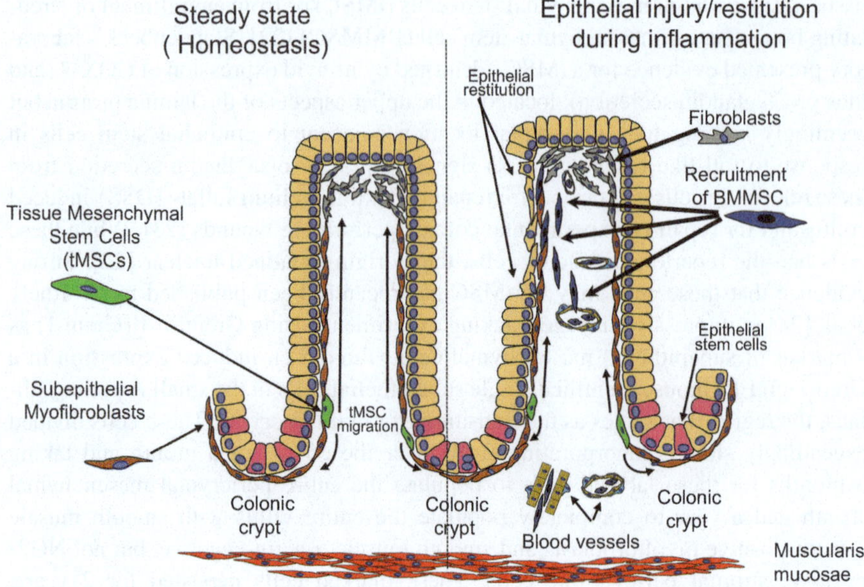

Fig. 7.1 Mesenchymal stem cell replacement of subepithelial myofibroblasts (stromal cells) during homeostasis (left) and following injury or damage from disease (right). Stromal cell replacement during homeostasis occurs by division of slow cycling tissue mesenchymal cells (tMSC) in a process that takes months to populate the top and bottom of the colonic crypts. After damage or disease, stromal cell replacement appears to be largely from recruitment of bone marrow-derived mesenchymal stem cells (BMMSC) which takes days or weeks. tMSC may also take part in stromal cells' replacement after damage

CD49c expression (Table 7.1). When these isolated CD146 colonic cells were place in differentiation media, they became osteocytes and differentiated less efficiently into chondrocytes, but not at all to adipocytes. Thus, we believe that, at least in normal intestinal mucosa, CD90$^+$ myofibroblasts/fibroblasts are restricted progenitor cells of mesenchymal stem cell origin. Functional differences between conventional MSC and intestinal stromal cells will no doubt be clarified over the coming years.

7.3 Immunosuppression by MSC

B7 Molecule-Mediated Suppression A fundamental property of MSC is their ability to alter the profile of dendritic cells, naive and effector T cells, and natural killer cells to induce an anti-inflammatory or tolerant phenotype [16]. While they express MHC class I constitutively, class II molecule expression must be induced. B7 co-stimulatory molecules CD80 (B7-1) and CD86 (B7-2) are robustly expressed by professional antigen-presenting cells (APCs) such as dendritic cells. These B7 ligands engage the T cell receptors CD28 (resting t cells) and CTLA-4 (activated T cells), although with a dramatically higher affinity (100- to 1000-fold higher) for

Table 7.1 Phenotypic comparison of BMMSC, gastrointestinal mucosal tMSC, and myofibroblasts

Marker	BMMSC	tMSC	Myofibroblasts	Description/function
CD4	nd	− [28]	nd	T cell coreceptor/interact with non-polymorphic regions of MHC class II and HIV protein gp120 [66]
CD10	+ [28]	+ [28]	nd	Metalloproteinase/development and cancer [67]
CD11b	nd	nd	− [11]	Integrin/adhesion, migration, phagocytosis, chemotaxis, cytotoxicity [68]
CD11c	−	nd	− [11]	Integrin/interacts with lipid A moiety of LPS [69]
CD13	++	+	±[a]	Aminopeptidase N/regulator of signals triggered by other receptors, apoptosis [70]
CD14	− [28]	− [28]	+[a]	Co-receptor for TLR4/implicated in LPS-induced skin fibroblast proliferation [71]
CD 24	− [28]	± [28]	+ [72]	Cell-cell and cell-matrix adhesion glycoprotein/facilitates metastasis [73]
CD29	++ [28]	+ [28]	+[a]	Integrin β1/adhesion receptor for ECM components/epithelial differentiation, development and tissue organization [74]
CD31	− [28]	− [28]	−[a]	Platelet endothelial cell adhesion molecule/T cell homeostasis, effector function and trafficking [75]
CD34	− [28]	− [28]	−[a]	Hematopoietic progenitor cell Ag/cell adhesion regulation [76]
CD44	++ [28]	++ [28]	++[a]	Glycoprotein, a hyaluronic acid receptor/regulates cell adhesion, proliferation, survival, migration, and differentiation [73]
CD45	+−[28]	− [28]	−[a]	Leukocyte common antigen, a transmembrane phosphatase/development and function of lymphocytes [77]
CD49a	++	++	nd	Integrin α1, heterodimerizes with the β1 subunit to form a cell surface receptor for collagen and laminin/adhesion [78]
CD49c	++ [28]	++ [28]	+[a]	Integrin α3, heterodimerizes with the β1 subunit/cell migration and adhesion, regulation of ECM components [79]
CD49d	++ [28]	++ [28]	++[a]	Integrin α4, heterodimerizes with the β1 subunit/interact with VCAM-1; cell adhesion [80]
CD54	++ [28]	++ [28]	++ [81]	Glycoprotein, also known as intercellular adhesion molecule 1, ICAM-1/cell adhesion [80]
CD80	± [32]	nd	− [11]	B7 family co-stimulator/interacts with CD28, CTLA-4, and PD-L1/regulation of T and macrophage activity [30, 57]
CD86	± [32]	nd	± [11]	B7 family co-stimulator/interacts with CD28, CTLA-4/regulation of T and macrophage activity [30, 57]
CD90	++	++	++	Activation-associated cell adhesion molecule (Thy1)/cell adhesion [82]

(continued)

Table 7.1 (continued)

Marker	BMMSC	tMSC	Myofibroblasts	Description/function
CD105	++ [28]	++	±	Also known as an endoglin, accessory receptor for TGF-β/implicated in angiogenesis and neovascularization [83]
CD146	++ [28]	++ [28]	++[a]	Cell adhesion molecule (CAM)/implicated in development, cell migration, mesenchymal stem cells differentiation, angiogenesis, immune responses [84]
CD166	++ [28]	++ [28]	+[a]	Activated leukocyte adhesion molecule (ALCAM)/bind to CD6; adhesion and T cell activity regulation [85]
HLA-ABC	++ [28]	++ [28]	++[a]	MHC class I molecules/MHC class I restricted Ag presentation to CD8[+] T cells [86]
HLA-DR	− [28]	++ [28]	+ [14]	MHC class II molecule/MHC class II restricted Ag presentation to CD4[+] T cells [86]
PD-L1	+ [87]	+ [87]	+ [14]	B7 family co-inhibitor (B7-H1)/interacts with PD-1 and CD80; regulation of T and macrophage activity [30, 57]
PD-L2	+ [87]	+ [87]	+ [14]	B7 family co-inhibitor (B7-DC)/interacts with PD-1; regulation of T and macrophage activity [30, 57]
B7-H2	nd	nd	+[a]	B7 family co-stimulator (ICOSL)/interacts with ICOS; activate T cell proliferation and induction of T17 type responses [88]
a-SMA	− [6]	+ [28]	+ [11]	Alpha-smooth muscle actin-2/involved in cell motility, structure, and contractile apparatus [6]
vimentin	+	+	++ [89]	Type III intermediate filament protein/major cytoskeletal component of mesenchymal cells; cell adhesion and endothelial sprouting [90]

[a]Unpublished data

nd non-determined, *Ag* antigen, *ECM* extracellular matrix

CTLA-4 [29–31]. Ligation of CD28 by CD80 and/or CD86 enhances T cell proliferation, intensifies pro-inflammatory cytokine secretion, and upregulates anti-apoptotic genes. MSC have low or negative expression of the positive B7 co-stimulatory molecules CD80 and CD86 but are reported to express high level of B7 inhibitory molecules PD-L1 and PD-L2 [32, 33]. These inhibitory molecules are critically involved in suppression of activated T lymphocyte proliferation, thus contributing to the maintenance of peripheral tolerance [33, 34]. PD-L1 is also reported to be implicated in MSC-mediated suppression of Th17 cell differentiation [35]. Importantly, there is evidence that PD-L1 expression on MSC may be responsible for suppression of autoreactive T cells in experimental autoimmune type 1 diabetes [36] and in inducing immune tolerance to cardiac allografts when given in combination with rapamycin [37]. Recent reports have demonstrated that PD-L1 is involved in the regulation of inflammatory Th17 [38] and immunosuppressive CD4[+] CD25[high] FoxP3[+] regulatory T cell (Treg) responses [39]. MSC have been shown to contribute to the regulation of the Th17/Treg balance and may repress mature Th17 cells in a PD-L1-dependent manner [35]. Taken together, these properties allow MSC to escape rapid immune rejection, and they establish the reason for the therapeutic value of MSC in the treatment of experimental and human immune-mediated diseases such as graft-versus-host disease, autoimmune encephalomyelitis, multiple sclerosis, type 1 diabetes mellitus, rheumatoid arthritis, systemic lupus erythematous, Crohn's disease, and cirrhosis, to name but a few [17, 40].

Soluble Mediator-Mediated Suppression The mechanisms that allow MSC-mediated immunosuppression were initially thought to occur only through secretion of soluble immune suppressors such as prostaglandin E2 (PGE2), indoleamine 2,3-dioxygenase (IDO, especially in human MSC), nitric oxide (NO, especially in murine MSC), and human leukocyte antigen (HLA)-G, as well as TGF-β, HGF, and hemoxygenase [17, 41–43]. Murine secretion of PGE_2 is upregulated by both interferon-γ (IFN-γ) and tumor necrosis factor-α (TNF-α), while IDO upregulation requires IFN-γ [32]. Therefore, soluble factor secretions of immunosuppressant molecules together with PD-L1-/PD-L2-mediated signaling are among the critical immunosuppressive mechanisms exerted by MSC which contribute to immune tolerance.

TLR Signaling Modulates Suppressive Properties of MSC While the biological and functional properties of murine and human MSCs differ, MSC from both species express toll-like receptors (TLRs) and NOD-like receptors (NLRs) NOD1 and NOD2 [18, 44]. TLRs and NLRs are known to trigger an innate immune response against microbial stimuli [45]. It has been suggested that stimulation of TLR3 has opposing effects from that occurring after activation of TLR4-mediated signaling [18]. TLR3 ligation by its putative ligand, dsRNA, results in an anti-inflammatory MSC phenotype with secretion of high levels of soluble immune suppressants, including IDO, PGE2, and TGF-B, and enhances MSC capacity to induce suppressive regulatory T (Treg) cells and M2 (suppressive) macrophages. Conversely, activation of TLR4 by its putative receptor, LPS, results in reduction of soluble suppressor secretion and in an increase in lymphocyte-recruiting chemokine (PIP-1a, MIP-1b, RANTES, CXCL9, and CXCL10) production [46]. However, Chen et al. [42] could not reproduce a differential suppressive effect of TLR 3 or TLR 4. In their experiments, ligation of neither TLR3 nor TLR4 affected the self-renewal, apoptosis, or expression of stem cell markers on MSC, while stimulation of TLR3 enhanced MSC differentiation into adipocytes and osteocytes, but activation of TLR4 signaling inhibited MSC differentiation. Thus, further investigation of TLR signaling in both MSC and MSC progeny, stromal cells, is necessary.

7.4 Immunosuppression by Intestinal Stromal Cells

B7 Molecule-Mediated Suppression While recent study of MSC immune function brought attention to the immunosuppressive potential of these cells [47, 48], initial studies of intestinal stromal myofibroblasts/fibroblasts (MFs) focused on their role in antigen presentation. In 2006 our group reported that human colonic MFs were among the major cell phenotypes in the normal human colonic lamina propria and were capable of presenting antigens in a MHC class II-dependent manner [13]. Expression of MHC class II was also observed on small intestinal and gastric MFs after stimulation with IFN-γ [12, 13]. In 2013, Owens et al. [49] demonstrated that, although somewhat limited when compared to dendritic cells, colonic MFs were able to uptake, phagocyte, and process *Salmonella typhimurium* antigens. Thus, MFs may possibly play a role of local APCs in the gastrointestinal mucosa.

MHC class II was shown to be involved in CD4$^+$ T cell proliferation induced by allogeneic and syngeneic MFs [11, 13, 50], but we observed that MFs isolated from healthy gut mucosa had a limited capacity to induce proliferation of naïve/resting CD4$^+$ T cells. Similar to the MSC, the limited capacity was thought to be due to constitutive absence of CD80 expression and low CD86 expression on MFs. These observations make it likely that CD86 mostly engages CTLA-4 on the activated effector T cells present in gut mucosa, and this engagement will contribute to the CTLA-4-mediated immunosuppression. Although further studies are required to understand the mechanisms and involvement of stromal cell CD86 expression in CTLA4-mediated suppression of activated T cells, a similar suppressor function has been proposed for immature dendritic cells which also have a low level of surface CD80 expression [51, 52].

Our finding of low levels of CD86 expression on normal human colonic MFs led us to hypothesize that these cells normally serve as "suppressors" of activated T cell responses in the healthy colon. MFs derived from normal colonic mucosa express strong basal level of PD-L1 and PD-L2 [14], and we found a similar robust expression of PD-L1 and PD-L2 in small intestinal and gastric MFs (unpublished data). As has been previously reported for MSC, PD-L1 and PD-L2 were found to be critically involved in MF-mediated suppression of the CD3-/CD28-activated CD4$^+$ T cell proliferation and IL-2 production [14].

Besides suppression of T cell proliferation, PD-L1 and PD-L2 are implicated in regulation of IFN-γ production by different immune cell subsets [53–56]. We demonstrated that PD-L1 is involved in the colonic MF-mediated suppression of the IFN-γ production by activated CD4$^+$T cells [57]. Recently we have observed that PD-L2 also contributes to MF-mediated suppression of both Th1 transcription factor T-bet expression and IFN-γ production by activated CD4$^+$T cells (unpublished data). Further studies are necessary to delineate the differences in the PD-L1- and PD-L2-mediated MF tolerogenic responses in the gut mucosa.

Suppression by Soluble Mediators While our laboratory has focused mostly on MF B7 molecule-mediated immunosuppression, similar to MSC, MFs in GI mucosa produce multiple soluble immunosuppressive cytokines, growth factors, and small metabolites (IL-10, IL-21, TGF-β, PGE$_2$, and IDO) [13, 50, 58, 59]. These molecules are known to contribute to the regulation of immune responses in the gut and are implicated in the regulation of Th1/Th17/Treg cell balance [5, 7]. Treg are especially important for maintaining gut mucosal tolerance [60]. We demonstrated that production of PGE$_2$ is critical to colonic MF-mediated induction of immunosuppressive Treg cells from naïve CD4$^+$CD45RA$^+$ T cells [50]. PD-L1 was minimally involved (contributing only ~10%) to colonic MF ability to induce Treg which appeared to be also dependent on MF expression of MHC class II [50]. Further studies will allow better understanding of the role of the soluble immunosuppressive molecule produced by MFs in their ability to promote tolerogenic responses.

TLR-Like Receptor-Mediated Modulation of Tolerogenic Responses The GI tract is populated by a resident and transitory microbiome. The continuous presence of normal physiological microflora in the GI lumen and mucosal surface provides a significant source for TLR and NLR ligands [45]. Signaling through these innate immunity receptors is thought to play a major role in orchestrating mucosal tolerogenic responses [57, 61]. Because MFs are located just beneath the basement membrane of the epithelial layers, are exposed to luminal ligands when epithelial tight junctions are leaky, express TLR 1–9 and NOD 1/NOD 2, and actively participate in wound repairs in GI mucosa [62], it is likely that MF-mediated immunosuppression is modulated by microbiota at least during wound healing process. Indeed, we recently demonstrated that stimulation of TLR2, TLR4, and TLR5 enhances the immunosuppressive capacity of normal colonic MFs via an increase in PD-L1 expression [57].

Myeloid differentiation factor 88 (MyD88) serves as an adaptor for the majority of TLRs (except TLR3) and is required for the initiation of intracellular signaling in response to the binding of a microbial ligand to TLR [63]. Using primary human MF cultures and fibroblast-specific MyD88 conditional knockout mice, we demonstrated that both basal- and TLR-induced levels of PD-L1 on MFs in the colonic mucosa depend on MyD88 [57]. TLR4-mediated upregulation of MF PD-L1 resulted in enhanced suppression of CD4$^+$ effector T cell proliferation and IFN-γ production. Taking into consideration the key role of PD-L1 in the negative regulation of Th1 and IFN-γ production, the TLR-mediated increase in PD-L1 expression by MFs might serve the function of tuning the immune balance between inflammation and tolerance in the colonic mucosa and would serve to protect the colonic mucosa against overt inflammatory responses toward otherwise innocuous microflora.

7.5 Summary and Future Challenges

We have highlighted the current knowledge supporting an emerging concept that mucosal CD90$^+$ stromal cells are partially differentiated MSC, and like MSC they are key participants in gut mucosa tolerogenic responses. While more extensive work is needed to understand the functional differences between MFs and MSCs, it is clear that MFs derived from normal GI mucosa preserve several MSC immunosuppressive functions through expression of common immunosuppressive molecules: PD-L1, PD-L2, PGE$_2$, IDO, and TGF-β. However, MFs acquire some specific innate immunogenic effector functions that are, perhaps, relevant to the specific organs/tissue [64]. The similarity and difference in molecule expression by BMMSC, tMSC, and MFs are summarized in Table 7.1. For instance, although less efficient than professional APCs, the intestinal MFs express MHC class II and are capable of the uptake, processing, and presenting of antigen to T cells [11, 49]. In contrast to the MSC, MFs express the positive co-stimulator CD86 (although this expression is limited) but strong constitutive expression of ICOSL (a.k.a. B7-H2) and B7-H3 (unpublished data) whose immune roles in MFs are unclear.

Multiple challenges must be overcome to better understand the role of these cells in the maintenance of health and in the development and progression of gastrointestinal inflammatory diseases. For example, although recently published evidence supports the mesenchymal origin of these cells, additional source of MFs has been described: epithelial to mesenchymal transition, endothelial to mesenchymal transition, and mucosal engraftment of circulating fibrocytes, presumably of hematopoietic origin. A better panel of MF-, tissue-, and lineage-specific markers is necessary to understand the stromal cell's role in chronic intestinal inflammatory diseases and cancers [5, 65]. Over the last decade, we have achieved some understanding on the role of the gastrointestinal MF in the regulation of the $CD4^+$ T cell responses. However, the role of these innate immune cells in the regulation of $CD8^+$ T cells, gamma/delta T cells, B cells, and professional APCs is unreported and will definitively be topics to clarify over the coming years.

Finally, recent published reports support the concept that mucosal stromal cells are innate immune cells contributing to the maintenance of the mucosal tolerance. A critical importance for stromal cells in inflammatory bowel disease and cancer has been suggested [5, 13, 50, 65]. Here, we have only discussed current knowledge of the immunological functions of stromal cells during homeostasis. However, we and others have observed that these cells appear to undergo hardwired phenotypic changes, switching from immunosuppressive to an inflammation-promoting phenotype at the chronic stage of the GI inflammatory diseases and cancers [13, 50]. Understanding these pathological processes will likely provide investigators with novel biomarkers and new therapeutic targets.

References

1. Powell DW, Mifflin RC, Valentich JD, Crowe SE, Saada JI, West AB. Myofibroblasts. I. Paracrine cells important in health and disease. Am J Phys. 1999;277:C1–9.
2. Powell DW, Mifflin RC, Valentich JD, Crowe SE, Saada JI, West AB. Myofibroblasts. II. Intestinal subepithelial myofibroblasts. Am J Phys. 1999;277:C183–201.
3. Furuya S, Furuya K. Subepithelial fibroblasts in intestinal villi: roles in intercellular communication. Int Rev Cytol. 2007;264:165–223.
4. De Wever O, Demetter P, Mareel M, Bracke M. Stromal myofibroblasts are drivers of invasive cancer growth. Int J Cancer. 2008;123:2229–38.
5. Powell DW, Pinchuk IV, Saada JI, Chen X, Mifflin RC. Mesenchymal cells of the intestinal lamina propria. Annu Rev Physiol. 2011;73:213–37.
6. Mifflin RC, Pinchuk IV, Saada JI, Powell DW. Intestinal myofibroblasts: targets for stem cell therapy. Am J Physiol Gastrointest Liver Physiol. 2011;300:G684–96.
7. Owens BM, Simmons A. Intestinal stromal cells in mucosal immunity and homeostasis. Mucosal Immunol. 2013;6:224–34.

8. Brahmer JR, Hammers H, Lipson EJ. Nivolumab: targeting PD-1 to bolster antitumor immunity. Future Oncol. 2015;11:1307–26.
9. Brahmer JR, Tykodi SS, Chow LQ, Hwu WJ, Topalian SL, Hwu P, Drake CG, Camacho LH, Kauh J, Odunsi K, Pitot HC, Hamid O, Bhatia S, Martins R, Eaton K, Chen S, Salay TM, Alaparthy S, Grosso JF, Korman AJ, Parker SM, Agrawal S, Goldberg SM, Pardoll DM, Gupta A, Wigginton JM. Safety and activity of anti-PD-L1 antibody in patients with advanced cancer. N Engl J Med. 2012;366:2455–65.
10. Lipson EJ, Sharfman WH, Drake CG, Wollner I, Taube JM, Anders RA, Xu H, Yao S, Pons A, Chen L, Pardoll DM, Brahmer JR, Topalian SL. Durable cancer regression off-treatment and effective reinduction therapy with an anti-PD-1 antibody. Clin Cancer Res. 2013;19:462–8.
11. Saada JI, Pinchuk IV, Barrera CA, Adegboyega PA, Suarez G, Mifflin RC, Di Mari JF, Reyes VE, Powell DW. Subepithelial myofibroblasts are novel nonprofessional APCs in the human colonic mucosa. J Immunol. 2006;177:5968–79.
12. Pinchuk IV, Beswick EJ, Saada JI, Suarez G, Winston J, Mifflin RC, Di Mari JF, Powell DW, Reyes VE. Monocyte chemoattractant protein-1 production by intestinal myofibroblasts in response to staphylococcal enterotoxin a: relevance to staphylococcal enterotoxigenic disease. J Immunol. 2007;178:8097–106.
13. Pinchuk IV, Morris KT, Nofchissey RA, Earley RB, Wu JY, Ma TY, Beswick EJ. Stromal cells induce Th17 during Helicobacter pylori infection and in the gastric tumor microenvironment. PLoS One. 2013;8:e53798.
14. Pinchuk IV, Saada JI, Beswick EJ, Boya G, Qiu SM, Mifflin RC, Raju GS, Reyes VE, Powell DW. PD-1 ligand expression by human colonic myofibroblasts/fibroblasts regulates CD4+ T-cell activity. Gastroenterology. 2008;135:1228–37.
15. Jones S, Horwood N, Cope A, Dazzi F. The antiproliferative effect of mesenchymal stem cells is a fundamental property shared by all stromal cells. J Immunol. 2007;179:2824–31.
16. Aggarwal S, Pittenger MF. Human mesenchymal stem cells modulate allogeneic immune cell responses. Blood. 2005;105:1815–22.
17. Singer NG, Caplan AI. Mesenchymal stem cells: mechanisms of inflammation. Annu Rev Pathol. 2011;6:457–78.
18. Bernardo ME, Fibbe WE. Mesenchymal stromal cells: sensors and switchers of inflammation. Cell Stem Cell. 2013;13:392–402.
19. Joseph NM, Mukouyama YS, Mosher JT, Jaegle M, Crone SA, Dormand EL, Lee KF, Meijer D, Anderson DJ, Morrison SJ. Neural crest stem cells undergo multilineage differentiation in developing peripheral nerves to generate endoneurial fibroblasts in addition to Schwann cells. Development. 2004;131:5599–612.
20. Wilm B, Ipenberg A, Hastie ND, Burch JB, Bader DM. The serosal mesothelium is a major source of smooth muscle cells of the gut vasculature. Development. 2005;132:5317–28.
21. Rinkevich Y, Mori T, Sahoo D, Xu PX, Bermingham JR Jr, Weissman IL. Identification and prospective isolation of a mesothelial precursor lineage giving rise to smooth muscle cells and fibroblasts for mammalian internal organs, and their vasculature. Nat Cell Biol. 2012;14:1251–60.
22. Powell DW, Saada JI. Mesenchymal stem cells and prostaglandins may be critical for intestinal wound repair. Gastroenterology. 2012;143:19–22.
23. Brown SL, Riehl TE, Walker MR, Geske MJ, Doherty JM, Stenson WF, Stappenbeck TS. Myd88-dependent positioning of Ptgs2-expressing stromal cells maintains colonic epithelial proliferation during injury. J Clin Invest. 2007;117:258–69.
24. Worthley DL, Churchill M, Compton JT, Tailor Y, Rao M, Si Y, Levin D, Schwartz MG, Uygur A, Hayakawa Y, Gross S, Renz BW, Setlik W, Martinez AN, Chen X, Nizami S, Lee HG, Kang HP, Caldwell JM, Asfaha S, Westphalen CB, Graham T, Jin G, Nagar K, Wang H, Kheirbek MA, Kolhe A, Carpenter J, Glaire M, Nair A, Renders S, Manieri N, Muthupalani S, Fox JG, Reichert M, Giraud AS, Schwabe RF, Pradere JP, Walton K, Prakash A, Gumucio D, Rustgi AK, Stappenbeck TS, Friedman RA, Gershon MD, Sims P, Grikscheit T, Lee FY, Karsenty G, Mukherjee S, Wang TC. Gremlin 1 identifies a skeletal stem cell with bone, cartilage, and reticular stromal potential. Cell. 2015;160:269–84.

25. Brittan M, Chance V, Elia G, Poulsom R, Alison MR, MacDonald TT, Wright NA. A regenerative role for bone marrow following experimental colitis: contribution to neovasculogenesis and myofibroblasts. Gastroenterology. 2005;128:1984–95.
26. Quante M, Tu SP, Tomita H, Gonda T, Wang SS, Takashi S, Baik GH, Shibata W, Diprete B, Betz KS, Friedman R, Varro A, Tycko B, Wang TC. Bone marrow-derived myofibroblasts contribute to the mesenchymal stem cell niche and promote tumor growth. Cancer Cell. 2011;19:257–72.
27. Worthley DL, Si Y, Quante M, Churchill M, Mukherjee S, Wang TC. Bone marrow cells as precursors of the tumor stroma. Exp Cell Res. 2013;319:1650–6.
28. Signore M, Cerio AM, Boe A, Pagliuca A, Zaottini V, Schiavoni I, Fedele G, Petti S, Navarra S, Ausiello CM, Pelosi E, Fatica A, Sorrentino A, Valtieri M. Identity and ranking of colonic mesenchymal stromal cells. J Cell Physiol. 2012;227:3291–300.
29. Tirapu I, Huarte E, Guiducci C, Arina A, Zaratiegui M, Murillo O, Gonzalez A, Berasain C, Berraondo P, Fortes P, Prieto J, Colombo MP, Chen L, Melero I. Low surface expression of B7-1 (CD80) is an immunoescape mechanism of colon carcinoma. Cancer Res. 2006;66:2442–50.
30. Seliger B, Marincola FM, Ferrone S, Abken H. The complex role of B7 molecules in tumor immunology. Trends Mol Med. 2008;14:550–9.
31. Romo-Tena J, Gomez-Martin D, Alcocer-Varela J. CTLA-4 and autoimmunity: new insights into the dual regulator of tolerance. Autoimmun Rev. 2013;12:1171–6.
32. English K, Barry FP, Field-Corbett CP, Mahon BP. IFN-gamma and TNF-alpha differentially regulate immunomodulation by murine mesenchymal stem cells. Immunol Lett. 2007;110:91–100.
33. Jang IK, Yoon HH, Yang MS, Lee JE, Lee DH, Lee MW, Kim DS, Park JE. B7-H1 inhibits T cell proliferation through MHC class II in human mesenchymal stem cells. Transplant Proc. 2014;46:1638–41.
34. Chen L. Co-inhibitory molecules of the B7-CD28 family in the control of T-cell immunity. Nat Rev Immunol. 2004;4:336–47.
35. Luz-Crawford P, Noel D, Fernandez X, Khoury M, Figueroa F, Carrion F, Jorgensen C, Djouad F. Mesenchymal stem cells repress Th17 molecular program through the PD-1 pathway. PLoS One. 2012;7:e45272.
36. Fiorina P, Jurewicz M, Augello A, Vergani A, Dada S, La Rosa S, Selig M, Godwin J, Law K, Placidi C, Smith RN, Capella C, Rodig S, Adra CN, Atkinson M, Sayegh MH, Abdi R. Immunomodulatory function of bone marrow-derived mesenchymal stem cells in experimental autoimmune type 1 diabetes. J Immunol. 2009;183:993–1004.
37. Wang H, Qi F, Dai X, Tian W, Liu T, Han H, Zhang B, Li H, Zhang Z, Du C. Requirement of B7-H1 in mesenchymal stem cells for immune tolerance to cardiac allografts in combination therapy with rapamycin. Transpl Immunol. 2014;31:65–74.
38. Dulos J, Carven GJ, van Boxtel SJ, Evers S, Driessen-Engels LJ, Hobo W, Gorecka MA, de Haan AF, Mulders P, Punt CJ, Jacobs JF, Schalken JA, Oosterwijk E, van Eenennaam H, Boots AM. PD-1 blockade augments Th1 and Th17 and suppresses Th2 responses in peripheral blood from patients with prostate and advanced melanoma cancer. J Immunother. 2012;35:169–78.
39. Francisco LM, Salinas VH, Brown KE, Vanguri VK, Freeman GJ, Kuchroo VK, Sharpe AH. PD-L1 regulates the development, maintenance, and function of induced regulatory T cells. J Exp Med. 2009;206:3015–29.
40. Cao W, Cao K, Cao J, Wang Y, Shi Y. Mesenchymal stem cells and adaptive immune responses. Immunol Lett. 2015;pii: S0165-2478(15)00101-7. https://doi.org/10.1016/j.imlet.2015.06.003. [Epub ahead of print].
41. Ghannam S, Bouffi C, Djouad F, Jorgensen C, Noel D. Immunosuppression by mesenchymal stem cells: mechanisms and clinical applications. Stem Cell Res Ther. 2010;1–7.
42. Hoogduijn MJ, Popp F, Verbeek R, Masoodi M, Nicolaou A, Baan C, Dahlke MH. The immunomodulatory properties of mesenchymal stem cells and their use for immunotherapy. Int Immunopharmacol. 2010;10:1496–500.
43. Stagg J, Galipeau J. Mechanisms of immune modulation by mesenchymal stromal cells and clinical translation. Curr Mol Med. 2013;13:856–67.

44. Kim HS, Shin TH, Yang SR, Seo MS, Kim DJ, Kang SK, Park JH, Kang KS. Implication of NOD1 and NOD2 for the differentiation of multipotent mesenchymal stem cells derived from human umbilical cord blood. PLoS One. 2010;5:e15369.
45. Wlodarska M, Kostic AD, Xavier RJ. An integrative view of microbiome-host interactions in inflammatory bowel diseases. Cell Host Microbe. 2015;17:577–91.
46. Chen X, Zhang ZY, Zhou H, Zhou GW. Characterization of mesenchymal stem cells under the stimulation of Toll-like receptor agonists. Develop Growth Differ. 2014;56:233–44.
47. Ishikura H, Dorf ME. Thymic stromal cells induce hapten-specific, genetically restricted effector suppressor cells in vivo. Immunobiology. 1990;182:11–21.
48. Krampera M, Glennie S, Dyson J, Scott D, Laylor R, Simpson E, Dazzi F. Bone marrow mesenchymal stem cells inhibit the response of naive and memory antigen-specific T cells to their cognate peptide. Blood. 2003;101:3722–9.
49. Owens BM, Steevels TA, Dudek M, Walcott D, Sun MY, Mayer A, Allan P, Simmons A. CD90(+) stromal cells are non-professional innate immune effectors of the human colonic mucosa. Front Immunol. 2013;4:307.
50. Pinchuk IV, Beswick EJ, Saada JI, Boya G, Schmitt D, Raju GS, Brenmoehl J, Rogler G, Reyes VE, Powell DW. Human colonic myofibroblasts promote expansion of CD4(+) CD25(high) Foxp3(+) regulatory T cells. Gastroenterology. 2011;79:2737–45.
51. Lohr J, Knoechel B, Jiang S, Sharpe AH, Abbas AK. The inhibitory function of B7 costimulators in T cell responses to foreign and self-antigens. Nat Immunol. 2003;4:664–9.
52. Probst HC, McCoy K, Okazaki T, Honjo T, van den Broek M. Resting dendritic cells induce peripheral CD8+ T cell tolerance through PD-1 and CTLA-4. Nat Immunol. 2005;6:280–6.
53. Alvarez IB, Pasquinelli V, Jurado JO, Abbate E, Musella RM, de la Barrera SS, Garcia VE. Role played by the programmed death-1-programmed death ligand pathway during innate immunity against Mycobacterium tuberculosis. J Infect Dis. 2010;202:524–32.
54. Chang KC, Burnham CA, Compton SM, Rasche DP, Mazuski RJ, McDonough JS, Unsinger J, Korman AJ, Green JM, Hotchkiss RS. Blockade of the negative co-stimulatory molecules PD-1 and CTLA-4 improves survival in primary and secondary fungal sepsis. Crit Care. 2013;17:R85.
55. Liang SC, Greenwald RJ, Latchman YE, Rosas L, Satoskar A, Freeman GJ, Sharpe AH. PD-L1 and PD-L2 have distinct roles in regulating host immunity to cutaneous leishmaniasis. Eur J Immunol. 2006;36:58–64.
56. McAlees JW, Lajoie S, Dienger K, Sproles AA, Richgels PK, Yang Y, Khodoun M, Azuma M, Yagita H, Fulkerson PC, Wills-Karp M, Lewkowich IP. Differential control of CD4(+) T-cell subsets by the PD-1/PD-L1 axis in a mouse model of allergic asthma. Eur J Immunol. 2015;45:1019–29.
57. Beswick EJ, Johnson JR, Saada JI, Humen M, House J, Dann S, Qiu S, Brasier AR, Powell DW, Reyes VE, Pinchuk IV. TLR4 activation enhances the PD-L1-mediated tolerogenic capacity of colonic CD90+ stromal cells. J Immunol. 2014;193:2218–29.
58. Ferdinande L, Demetter P, Perez-Novo C, Waeytens A, Taildeman J, Rottiers I, Rottiers P, De Vos M, Cuvelier CA. Inflamed intestinal mucosa features a specific epithelial expression pattern of indoleamine 2,3-dioxygenase. Int J Immunopathol Pharmacol. 2008;21:289–95.
59. Ina K, Kusugami K, Kawano Y, Nishiwaki T, Wen ZH, Musso A, West GA, Ohta M, Goto H, Fiocchi C. Intestinal fibroblast-derived IL-10 increases survival of mucosal T cells by inhibiting growth factor deprivation- and fas-mediated apoptosis. J Immunol. 2005;175:2000–9.
60. Allez M, Mayer L. Regulatory T cells: peace keepers in the gut. Inflamm Bowel Dis. 2004;10:666–76.
61. Wang S, Charbonnier LM, Noval Rivas M, Georgiev P, Li N, Gerber G, Bry L, Chatila TA. MyD88 adaptor-dependent microbial sensing by regulatory T cells promotes mucosal tolerance and enforces commensalism. Immunity. 2015;43:289–303.
62. Otte JM, Rosenberg IM, Podolsky DK. Intestinal myofibroblasts in innate immune responses of the intestine. Gastroenterology. 2003;124:1866–78.
63. Deguine J, Barton GM. MyD88: a central player in innate immune signaling. F1000Prime Rep. 2014;6:97.

64. Owens BM. Inflammation, innate immunity, and the intestinal stromal cell niche: opportunities and challenges. Front Immunol. 2015;6:319.
65. Rieder F, Fiocchi C. Intestinal fibrosis in inflammatory bowel disease: progress in basic and clinical science. Curr Opin Gastroenterol. 2008;24:462–8.
66. Dumas F, Preira P, Salome L. Membrane organization of virus and target cell plays a role in HIV entry. Biochimie. 2014;107(Pt A):22–7.
67. Maguer-Satta V, Besancon R, Bachelard-Cascales E. Concise review: neutral endopeptidase (CD10): a multifaceted environment actor in stem cells, physiological mechanisms, and cancer. Stem Cells. 2011;29:389–96.
68. Zhou H, Liao J, Aloor J, Nie H, Wilson BC, Fessler MB, Gao HM, Hong JS. CD11b/CD18 (Mac-1) is a novel surface receptor for extracellular double-stranded RNA to mediate cellular inflammatory responses. J Immunol. 2013;190:115–25.
69. Flaherty SF, Golenbock DT, Milham FH, Ingalls RR. CD11/CD18 leukocyte integrins: new signaling receptors for bacterial endotoxin. J Surg Res. 1997;73:85–9.
70. Mina-Osorio P. The moonlighting enzyme CD13: old and new functions to target. Trends Mol Med. 2008;14:361–71.
71. Yang H, Li J, Wang Y, Hu Q. Effects of CD14 and TLR4 on LPS-mediated normal human skin fibroblast proliferation. Int J Clin Exp Med. 2015;8:2267–72.
72. Lang M, Schlechtweg M, Kellermeier S, Brenmoehl J, Falk W, Scholmerich J, Herfarth H, Rogler G, Hausmann M. Gene expression profiles of mucosal fibroblasts from strictured and nonstrictured areas of patients with Crohn's disease. Inflamm Bowel Dis. 2009;15:212–23.
73. Jaggupilli A, Elkord E. Significance of CD44 and CD24 as cancer stem cell markers: an enduring ambiguity. Clin Dev Immunol. 2012;2012:708036.
74. Yeh YC, Lin HH, Tang MJ. A tale of two collagen receptors, integrin beta1 and discoidin domain receptor 1, in epithelial cell differentiation. Am J Physiol Cell Physiol. 2012;303:C1207–17.
75. Marelli-Berg FM, Clement M, Mauro C, Caligiuri G. An immunologist's guide to CD31 function in T-cells. J Cell Sci. 2013;126:2343–52.
76. Nielsen JS, McNagny KM. Novel functions of the CD34 family. J Cell Sci. 2008;121:3683–92.
77. Thiel N, Zischke J, Elbasani E, Kay-Fedorov P, Messerle M. Viral interference with functions of the cellular receptor tyrosine phosphatase CD45. Viruses. 2015;7:1540–57.
78. Gardner H. Integrin alpha1beta1. Adv Exp Med Biol. 2014;819:21–39.
79. Subbaram S, Dipersio CM. Integrin alpha3beta1 as a breast cancer target. Expert Opin Ther Targets. 2011;15:1197–210.
80. Mitroulis I, Alexaki VI, Kourtzelis I, Ziogas A, Hajishengallis G, Chavakis T. Leukocyte integrins: role in leukocyte recruitment and as therapeutic targets in inflammatory disease. Pharmacol Ther. 2015;147:123–35.
81. Musso A, Condon TP, West GA, De La Motte C, Strong SA, Levine AD, Bennett CF, Fiocchi C. Regulation of ICAM-1-mediated fibroblast-T cell reciprocal interaction: implications for modulation of gut inflammation. Gastroenterology. 1999;117:546–56.
82. Saalbach A, Wetzel A, Haustein UF, Sticherling M, Simon JC, Anderegg U. Interaction of human Thy-1 (CD 90) with the integrin alphavbeta3 (CD51/CD61): an important mechanism mediating melanoma cell adhesion to activated endothelium. Oncogene. 2005;24:4710–20.
83. Nassiri F, Cusimano MD, Scheithauer BW, Rotondo F, Fazio A, Yousef GM, Syro LV, Kovacs K, Lloyd RV. Endoglin (CD105): a review of its role in angiogenesis and tumor diagnosis, progression and therapy. Anticancer Res. 2011;31:2283–90.
84. Wang Z, Yan X. CD146, a multi-functional molecule beyond adhesion. Cancer Lett. 2013;330:150–62.
85. Chappell PE, Garner LI, Yan J, Metcalfe C, Hatherley D, Johnson S, Robinson CV, Lea SM, Brown MH. Structures of CD6 and its ligand CD166 give insight into their interaction. Structure. 2015;23:1426–36.
86. Nakayama M. Antigen presentation by MHC-dressed cells. Front Immunol. 2014;5:672.

87. Augello A, Tasso R, Negrini SM, Amateis A, Indiveri F, Cancedda R, Pennesi G. Bone marrow mesenchymal progenitor cells inhibit lymphocyte proliferation by activation of the programmed death 1 pathway. Eur J Immunol. 2005;35:1482–90.
88. Lina TT, Pinchuk IV, House J, Yamaoka Y, Graham DY, Beswick EJ, Reyes VE. CagA-dependent downregulation of B7-H2 expression on gastric mucosa and inhibition of Th17 responses during Helicobacter pylori infection. J Immunol. 2013;191:3838–46.
89. Adegboyega PA, Mifflin RC, DiMari JF, Saada JI, Powell DW. Immunohistochemical study of myofibroblasts in normal colonic mucosa, hyperplastic polyps, and adenomatous colorectal polyps. Arch Pathol Lab Med. 2002;126:829–36.
90. Dave JM, Bayless KJ. Vimentin as an integral regulator of cell adhesion and endothelial sprouting. Microcirculation. 2014;21:333–44.

Chapter 8
Novel Models to Study Stromal Cell-Leukocyte Interactions in Health and Disease

Mattias Svensson and Puran Chen

Abstract To study human immunology in general and stromal immunology in particular, it is highly motivated to move from monolayers to 3D cultures, such as organotypic models, that better mimic the function of living tissue. These models can potentially contain most if not all cell types present in tissues, in combination with different extracellular matrix components that can critically affect cell phenotype. Besides their well-established use in studies of tissue-specific cells, such as epithelial cells, endothelial cells and stromal fibroblasts in combination with extracellular components, these models have also been shown to be valuable to study how tissue participates in the regulation of leukocyte differentiation and function. Organotypic models with leukocytes represent novel powerful tools to study human stromal immunology and mechanisms involved in the regulation of leukocyte functions and inflammatory processes in human health and disease. In particular, these models are robust, long-lived and reproducible and allow monitoring of disease progression in real time, as well as the mixing of cellular constituents from healthy and pathological tissues. These models are also easy to manipulate, either genetically or by adding external stimulants, such as cytokines and pathogens, to mimic pathological conditions. It is thus not surprising that these models are proposed to be useful in toxicology screening assays, evaluating therapeutic efficacy of drugs and antibiotics, as well as in personalized medicine. Within this chapter, the most recent developments in creating organotypic models for the purpose of study of human leukocyte and stromal cell interactions, in health and disease, will be discussed, in particular focusing on live imaging. Special emphasis will be given on an organotypic model resembling human lung and its usefulness in studying the fine control of physiological and pathological processes in human health and disease. Using these models in studies on human stromal cell and leukocyte interactions will likely help identifying novel disease traits and may point out new potential targets to monitor and treat human diseases.

M. Svensson (✉) · P. Chen
Center for Infectious Medicine, F59, Department of Medicine, Karolinska Institutet,
Karolinska University Hospital, Huddinge, Stockholm, Sweden
e-mail: mattias.svensson@ki.se

© Springer International Publishing AG, part of Springer Nature 2018
B. M.J Owens, M. A. Lakins (eds.), *Stromal Immunology*, Advances in Experimental
Medicine and Biology 1060, https://doi.org/10.1007/978-3-319-78127-3_8

Keywords Tissue microenvironment · Tissue inflammation · Cell migration · Live imaging · Organotypic models

8.1 Introduction

Host responses to pathogens and inflammatory reactions in tissue are highly dynamic processes dependent on the cooperation between leukocytes, tissue-specific cells and extracellular matrix components [1–3]. Thus, complexities of cell-cell interactions between the same and different types of cells, as well as cell-matrix interactions, must be considered when performing studies of, for example, host-pathogen interactions, tumorigenesis and inflammatory reactions in tissue. As the interest of tissue-specific cells including epithelial cells, stromal fibroblasts and endothelial cells, as well as extracellular matrix proteins, and the interactions with leukocytes and pathogens have grown increasingly, a variety of approaches have been developed. Approaches span the spectrum from two-dimensional (2D) cultures with mixed cell populations to living organisms, with the study systems along the spectrum having their respective strengths and weaknesses. The 2D culture systems are easy to use and highly reproducible, but the representation of intact tissue is poor. Furthermore, many cells isolated from tissue biopsies are terminally differentiated, short-lived cells that generally have lower metabolic capacity than actively growing cells in vivo or in vitro [4]. These monolayer systems are also limited by the lack of polarized cell phenotype and lack a large number of cell-cell contacts, which affects their function and response to external stimuli [5]. Also, constitutive as well as inducible expression of proteins in these cells may be highly variable due to donor variability. Thus, moving from monolayers to three-dimensional (3D) cultures that better mimic the function of living tissue is motivated [6]. Intact tissues represented by tissue explants from, for example, skin punch biopsies and bronchoscopies [7] represent the real set of complex constituents and behaviours but have some limitations on how they can be manipulated and monitored in real time, particularly in humans. In contrast, advanced technologies, such as intravital microscopy, have been developed to study cell-cell and cell-matrix interactions during homeostasis and inflammation, as well as host-pathogen interactions in 3D environments in vivo, but these approaches are often limited to animal models [8, 9]. Organotypic culture models can offer a balance of strengths and weaknesses, and although they do not contain all of the cell types present in tissues, they often include endothelium or epithelium in combination with stromal cells and extracellular matrix, which can critically affect cell phenotype and function [10–14]. To date, highly reproducible 3D organotypic models of oral and lung mucosa as well as skin epidermal/dermal tissue among others have been developed. These 3D tissue models are typically engineered using cell lines or directly isolated primary tissue-specific cells. There are also examples of approaches where human 3D skin equivalents and 3D vascular networks have been reconstituted from induced pluripotent stem cells [15, 16]. Altogether, the further development and use of such approaches to build 'immunocompetent' human tissue models will allow researchers to perform immunological-based assays to study interactions between human tissue-specific cells and leukocytes, as well as mechanisms regulating physiologic and pathologic immunological events in live tissue. To date, a limited number of such 3D model tissue cells have been described, including models recapitulating the human lung [17], skin [18] and intestinal tissues [19]. Approaches aimed at further developing in vitro human 3D models that allow tissue-specific cells to be combined with leukocytes are highly warranted and will provide useful tools to study human stromal immunology and mechanisms involved in the regulation of leukocyte functions and inflammatory processes in the microenvironment.

8.2 Recapitulating Human Tissue in Three-Dimensional Models

To build appropriate 3D models, there are various aspects that need to be taken into account. This includes the species and target tissue to be modelled and what specific processes that will be studied. These and other aspects should determine which cell types and analytical methods may be used to answer the specific questions of interest.

8.2.1 Human 3D Tissue Models

Different species' immune systems are to some extent identical, but there are also important differences, all of which may be important for specific disease processes which may be poorly imitated in non-related species. For example, there are interspecies variations in bacterial binding surface receptors, important for infection [20, 21]. Other examples are certain chemokines that are only found in humans and contribute to inflammatory processes by recruiting leukocytes to inflammatory and/or infected tissues, as well as directing leukocytes in the target tissue [22–24]. Recapitulating human tissue in 3D tissue models can therefore serve as important tool to recapitulate certain aspects of human health and disease, for example, to delineate mechanisms of tissue pathology associated with aberrant inflammatory reactions to infections, cancer and chronic inflammation. Indeed, there is evidence demonstrating that adhesion and migration are markedly different in cells within 3D environments [13]. Three-dimensional lung tissue models of normal airway mucosa have proven to be informative for analysing the behaviour of epithelial cells, in particular [12, 25, 26]. Altogether, the multicellular assembly provides secreted factors and multiple cell-to-cell communications within the tissue microenvironment that is likely to play an essential role in regulating leukocyte function by stromal cells and vice versa [27] in different pathological settings (Fig. 8.1). In addition to the 'large-scale' approaches described herein, the application of microtechnology, as demonstrated for human liver tissue (hepatocytes and fibroblasts) [28], may allow the engineering of microscale tissue subunits for high-throughput experiments, such as toxicology screenings, and testing of therapeutic efficacy of novel immunomodulatory drugs in various tissues.

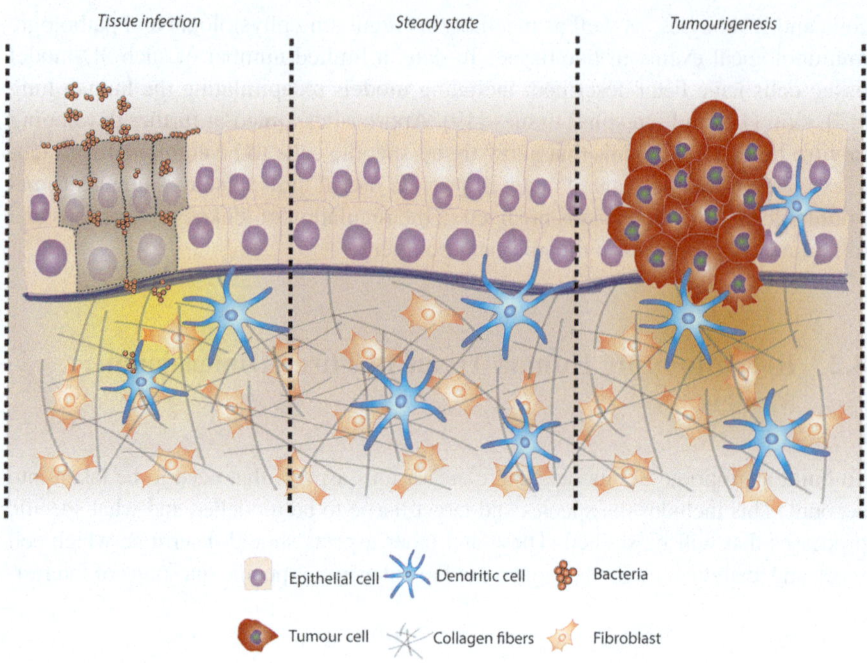

Fig. 8.1 Illustration of established model. Schematic drawing that illustrates the composition of the human 3D lung tissue model and its usefulness in various research areas in which the stroma-leukocyte interactions will be explored

8.2.2 The Stroma and Other Tissue-Specific Components

The stroma consists principally of fibroblasts and extracellular matrix proteins (ECM), which vary in phenotype depending on the tissue. The fibroblast component is not only critical in promoting survival, remodelling and deposition of matrix components but also ensures the maintenance of elasticity to lung tissue [10, 11]. Fibroblasts actively interact with the adjacent epithelial layer and have a key role in inflammation and tissue repair [29]. Several models with a physiologically relevant fibroblast matrix layer and a stratified epithelium that better recapitulate human tissue have been developed [12, 19, 30, 31]. Under such conditions, fibroblasts proliferate slowly, and their extracellular matrix is likely to provide better conditions for growth and differentiation of epithelial cell barriers in vitro compared with artificial gels or membranes, possibly via the release of growth and survival factors [11, 32]. The selection of scaffolds probably also influences cell functionality, at least initially, before the cells themselves begin to produce ECMs and build structures such as basement membranes, essential for tissue function. In this context, both biological and nonbiological compounds are used. Biological material includes various types of collagen type I and type IV, or Matrigel®, which is a complex protein mixture. More recently, nonbiological scaffolds based on nanotechnology and the synthesis of fibre structures using polymers have been developed. Although different type of scaffolds may be used, it is important that the cellular constituents retain their phenotypes and functions that best mimic the tissue of origin. Fibroblasts are relatively easy to derive from human tissue biopsies and kept in culture without immortalization, but it may be more difficult to establish in vitro cultures with, for example, primary lung epithelial cells or skin keratinocytes. An alternative approach used by many investigators is to use in vitro immortalized cells of interest; however, it is recommended to verify that the cells used have not lost important features. Certain epithelial cells should, for example, have the capacity to stratify and form functional epithelial barriers, including the production of adherence and tight junction proteins. Carcinoma-derived cell lines should preferably be avoided, unless certain functionalities associated with cancer are studied, as these cells can display abnormal organ-/tissue-specific functions. For example, the bronchial 16HBE14o- (16HBE) cell line is derived from a healthy individual and form proper epithelial barriers, while the adenocarcinoma cell line A549 form aberrant tight junctions and epithelial barriers [33]. However, such cell types are often reproducible and inexpensive, but whenever possible, it is advisable to use primary cell or immortalized cells derived from a healthy individual. Primary human endothelial cells also seem to grow in organotypic models and retain their original function to invade and sprout in various matrices [34, 35]. Often human umbilical vein endothelial cells are used; however, accumulating evidence suggests that endothelial cells (ECs) display significant heterogeneity across tissue types, playing an important role in tissue-specific regeneration and homeostasis. It has therefore been suggested that endothelial cell can be generated from human pluripotent stem cells and specific differentiation protocols may be directed towards tissue-specific fates [36]. Although endothelial cells have been cultured mainly in the context of different cellular matrices, the introduction of endothelial cells into the already existing organotypic models composed of stromal cells, epithelial barriers and leukocytes is the future goal. In this respect, the studies on bioartificial lung engineering using different types of synthetic scaffold material may provide useful information into how to further develop organotypic models to introduce essential structures for different types of cellular constituents [37].

8.2.3 *The Leukocyte Component*

Tissue macrophages and dendritic cells (DC) are phagocytic innate leukocytes that respond directly to pathogens or via environmental signals and in turn modulate other cells locally or after migration to secondary lymphoid organs. The use of leukocytes in organotypic models has so far mainly included the implantation of phagocytic cells [38–41]. In 3D models of human lung, human monocytes from peripheral blood as well as macrophages and dendritic cells generated in vitro from blood monocytes using M-CSF or GM-CSF and IL-4, respectively, have been used for implantation [40, 41]. In addition, a 3D model of human skin has successfully been used to implant Langerhans cells/dendritic cell precursors, which were generated by culturing haematopoietic progenitor cells cultured in GM-CSF, TGFβ1 and TNF prior to implantation [18]. The use of leukocytes with lymphoid origin (i.e. T cells and natural killer cells) is more challenging, as HLA mismatches can induce unwelcomed lymphocyte activation and in worst cases cytotoxic responses against the cells building up the model tissue. This may be overcome by establishing autologous tissue models in which the leukocytes and tissue-specific cells are derived from the same donor or with a matched HLA haplotype. Thus, it should be possible to build organotypic models with a variety of cell types in different combinations depending on the specific questions to be answered.

8.2.4   Tissue Model Set-Up

Human organotypic models or 3D tissue models with leukocytes can preferably be built in 3.0 μm transwell inserts placed in six-well plates based on a protocol for human oral mucosal models [19]. The procedure of establishing a model tissue takes approximately 2 to 3 weeks (Fig. 8.2) and includes cell expansion, culturing of the cells in the model and air exposure of the tissue model. Using this approach, we have successfully established three different organotypic models recapitulating human lung and oral or skin tissue, in which human phagocyte cells can be implanted. The established standardized human 3D organotypic lung model, recapitulating human lung described below, uses the 'normal' bronchial epithelial cell line 16HBE immortalized with the SV40 large T antigen, the MRC-5 fibroblasts derived from foetus lung tissue and human monocyte-derived DC [40]. From day 3 of air exposure, tissue models are used for immunohistological, immunofluorescence as well as live-imaging microscopy analyses. At this stage the model recapitulates key anatomical and functional features of lung mucosal tissue, including deposition of extracellular matrix proteins and the formation of tight junctions and adherence junctions. Once the epithelial cell layer is formed, epithelial cells proliferation decreases, likely due to the presence of fibroblasts [42, 43], and the model can be maintained in culture for long-term experiments (>two month). The intactness of the epithelial barrier is confirmed with transepithelial resistance measurements. The deposition of structural framework proteins, such as the adherence and tight junction proteins, E-cadherin and claudin-1, respectively, is visualized in the appropriate compartment of the model. By enzymatically digesting the established tissue models, DC survival and phenotype of DC (CD45, HLA-DR and CD1a) can be verified [39]. The confocal microscopy analyses of lung models also revealed that DC were distributed mainly in the subepithelial layer, and their survival was confirmed also in live tissue models labelling DC with a cell tracker dye prior to implantation and using live-imaging confocal or multiphoton microscopy analysis (Fig. 8.3). Notably, the implanted DC can survive for at least 1 month without the addition of external growth factors. Whether the DC retain or change their phenotype and functionality over such long culturing periods remains to be investigated in more detail.

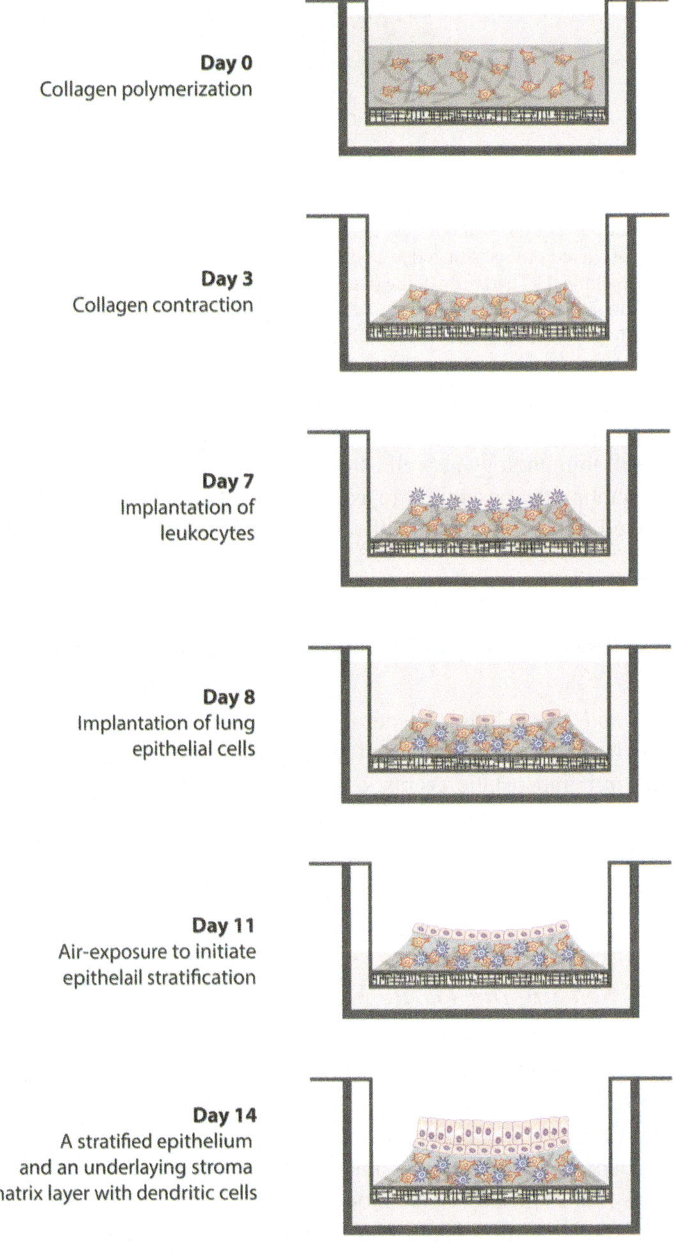

Fig. 8.2 Tissue model set-up. Schematic drawing of the model set-up in a six-well insert from day 1 to the time of live imaging, totally 14–16 days. Day 1: Culture MRC-5 fibroblasts in bovine collagen type I for 7 days in a six-well insert, and by day 3 the collagen should have contracted. Day 7: Add monocyte-derived DC on top of the fibroblast-collagen layer. Day 8: Seed 16HBE epithelial cells on top of the DC-fibroblast layer and submerge the culture in medium for 3 days. Day 11: Culture the model in an air-liquid interface for 3–5 days. Day 14–16: Stimulate the models for live-imaging experiments

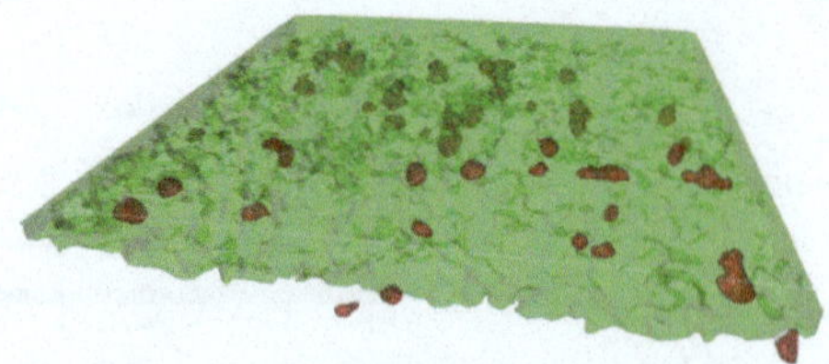

Fig. 8.3 3D rendering of tissue model. A representative 3D projection from live-imaging microscopy showing DC in red/orange situated just beneath the epithelium in green (fibroblasts excluded). At the time of imaging, DC had been in the tissue model for 1 month. Epithelial cells constitutively express green fluorescent protein, and DC were labelled prior to implantation with a cell tracker dye (PK-26). Images were acquired with a Nikon A1R confocal microscope

Nevertheless, combining DC and other leukocytes with tissue-specific stromal cells in organotypic models will most likely provide deeper understanding of fundamental stromal cell immunology as well as immunologic traits predisposing to immune activation, enabling the possibility to assess host-pathogen interactions and tumorigenesis, as well as the efficacy of adjuvants and immunomodulatory regimens.

8.3 Investigating Stromal Cell-Leukocyte Interactions

Macrophage and DC belong to heterogeneous populations of widely distributed phagocyte cells, with overlapping and unique functions, and both cell types play central roles in homoeostatic events, as well as the initiation and orchestration of immune responses [44–46]. Within the lung, DC mainly associate with the epithelial layer, and there is evidence of DC regulation by the epithelium, and that epithelial dysfunction leads to overzealous immune cell activation [47, 48].

8.3.1 DC Sensing the Tissue Microenvironment

Further evidence that DC are regulated by the lung tissue microenvironment comes the observations that DC implanted in the lung model produce CCL18, a chemokine that is constitutively expressed in lung at steady state and is elevated in several human disorders, including various malignancies and inflammatory lung, skin and joint diseases [22, 24]. Lung tissue models with DC abundantly expressed CCL18 mRNA and protein, while CCL18 expression was undetectable in the lung tissue model without DC [40]. In contrast, CCL17 and CCL22, which are chemokines that are barely detectable in peripheral tissues at steady state and that are induced during inflammatory reactions and associated to human skin and pulmonary inflammatory diseases [49, 50], were suppressed in tissue model DC compared to DC cultured in GM-CSF and IL-4. These findings support the importance of studying DC biology in more physiological relevant milieus rather than in monolayers on plastic surfaces.

These investigations also revealed that the complexity of tissue generated products inducing chemokine production by DC. This was shown by the fact that soluble components secreted from the 3D lung model, rather than from monolayers of fibroblast or epithelial cells, induced CCL18 production in DC.

8.3.2 3D Live Imaging of the Tissue Microenvironment

The organotypic model was further developed to enable live imaging and quantification of phagocyte cell (DC or macrophages) migratory behaviour in lung epithelial tissue. For live-imaging experiments, the 16HBE were transduced to express fluorescent proteins, while DC were labelled with a cell tracker dye prior to implantation. Live-imaging experiments were performed using an inverted confocal microscope that enables sequential analysis of six models at a time, and DC or macrophages were imaged at a depth of up to approximately 150 µm from the epithelial surface. Confocal image analysis revealed a well-defined stratified layer of epithelial cells well separated from the underlying collagen matrix of fibroblasts. Although both DC and macrophages were found on both sides of the boundary separating the fibroblast matrix layer and the epithelial layer, several DC and macrophages were identified interacting closely with the stroma fibroblasts (Fig. 8.4 and data not shown). In addition to defining phagocyte cell distribution relative to stromal cells and epithelial cells, the live-imaging set-up also allows quantitative analysis of phagocyte cell migration by tracking individual cells in the tissue at steady state and in response to stimulation (Fig. 8.5). Together, these data indicate that the organotypic model of the human lung is well suited for studying the interplay between stromal cells and phagocyte cells, how this might impact on phagocyte cell distribution and migration and the induction of inflammatory responses [51].

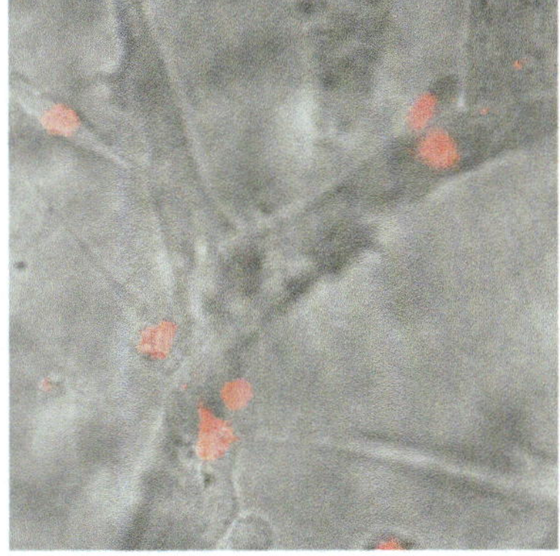

Fig. 8.4 Leukocyte interaction with stromal cells. A representative image of the lung model with dendritic cells (red) closely associated with the reticular stromal network

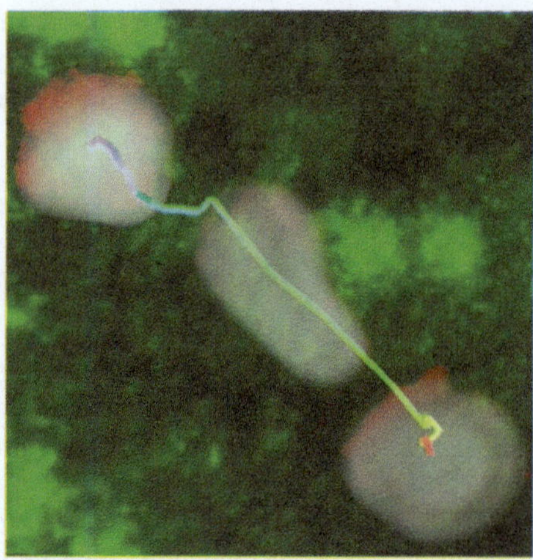

Fig. 8.5 Illustration of live-imaging 3D tracking. A representative image showing migration of one macrophage (red) in the organotypic human lung model. Epithelial cells constitutively express green florescent protein, and macrophages were labelled prior to implantation with a cell tracker dye (PK-26). Images were acquired with a Nikon A1R confocal microscope. To quantitatively investigate macrophage migratory behaviour in response to various stimuli, live-imaging fluorescence microscopy data were analysed with Imaris Surface tracking algorithm and Imaris Vantage. The line represents the centre of the macrophage in a stimulated (bacterial culture supernatant) model over time, where violet colouring of the line represents the position at 0 hours and red colouring of the line represents the position 14 hours post stimulation

8.4 Three-Dimensional Tissue Models to Understand Human Infection, Inflammation and Pathology

Using human tissue models has several advantages. For example, many important human pathogens (e.g. *Mycobacterium tuberculosis (Mtb), group A streptococcus (GAS) and Staphylococcus aureus*) induce species-specific responses [25, 52]. Along these lines, we have successfully established skin organotypic models suitable for GAS infections and lung organotypic models for Mtb or *S. aureus* infections.

8.4.1 Bacterial Infection of the Lung Model

To model human tuberculosis (TB), infections are usually performed with human blood cells in monolayer cultures [53, 54], which poorly represents the real course of the disease. As an alternative approach, monocyte-derived macrophages were infected and implanted in the lung tissue model described above. Histological analysis revealed macrophage-associated acid-fast bacteria in the tissues 7 days post implantation of macrophages. The association of Mtb with macrophages was further assessed by introducing macrophages infected with GFP-expressing Mtb and subsequent immunofluorescence-based detection of the macrophages at day 7 post implantation. In further support of the 3D lung tissue model's suitability for studies of human TB, fluorescently labelled monocytes migrated towards infected macrophages and clustered [41], defining initiation of granuloma formation with Mtb-infected macrophages [55]. Also, confocal microscopy analyses revealed that infection with the virulent H37Rv, but not the nonvirulent strain H37Ra or BCG, induced clustering of macrophages [41]. The virulence RD1 region of Mtb is required for granuloma formation [55–57], and using a ΔRD1 strain to infect the lung model revealed an inability to induce clustering of macrophages. Similarly, an Mtb strain with a deletion in the virulence gene ESTAT-6 was unable to induce clustering of phagocytic cells at site of infection, further underlining the suitability of the lung models for studies on human TB.

In addition, exposure of the lung model to *S. aureus*-derived toxins revealed tissue-destructive events (Fig. 8.6) and demonstrates that usefulness of the model to delineate mechanisms of inflammation and tissue pathology, including disruption of the lung epithelium [58]. In line with previous studies, the loss of tissue integrity was likely mediated via the activation of a disintegrin and metalloprotease 10 (ADAM10), which activities include proteolytic cleavage of the adherence junction protein E-cadherin [59, 60]. Furthermore, the relative contribution of stromal cells and phagocyte cells to the local inflammatory reaction and exaggerated pathology is currently under investigation.

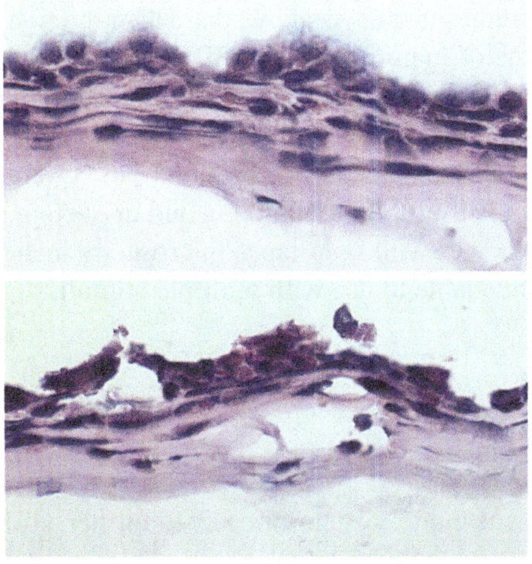

Fig. 8.6 Illustration of tissue bacterial toxin-induced pathology. For cryosectioning, lung tissue was treated with 2.0 M of sucrose for 1 hour before embedding in optimum cutting temperature compound (Sakura Finetek) followed by freezing in liquid nitrogen and stored at −80 °C until used. Eight μm cryosections were cut onto diagnostic microscope glass slides (Thermo Scientific, Waltham, MA) using a MICROM cryostat HM 560 MV (Carl Zeiss, Jena, Germany) and fixed in 2% freshly prepared formaldehyde in PBS for 15 minutes at RT or in ice-cold acetone for 10 minutes at −20 °C. Haematoxylin/eosin staining of cryosections of lung tissue, unstimulated (top) or stimulated (bottom) for 24 hours with *S. aureus* bacterial culture supernatant. The sections were stained for 15 sec in Mayer's haematoxylin and counterstained for 2 minutes in eosin

8.4.2   Inflammation and Phagocyte Cell Migration

By responding to toll-like receptor (TLR) ligands (PAMPs) and danger signals (DAMPs), DC and macrophages determine how immune responses to pathogens are initiated and completed, as well as the magnitude of inflammation [61–63]. Both DC and macrophages partake in orchestrating immune responses to pathogens by sensing pathogens and producing cytokines and chemokines that are important in the activation and recruitment of other inflammatory cells [64]. Although TLR ligands can induce DC maturation directly, other cells in the tissue such as epithelial cells and fibroblasts also express TLR and release cytokines that can contribute to induction of DC responses, such as cytokine production and increased cell motility. To explore the effect of TLR ligands on DC migration in a 3D microenvironment, the organotypic lung model was stimulated with TLR-4 and TLR-1/2 ligands. In addition, recombinant human CCL2, a chemokine known to induce DC migration [65], as well as DC maturation [66], was administered to the lung tissue model. Comparing the position of each DC in relation to the epithelial cell layer 4 hours post stimulation revealed that TLR-1/2 and CCL2, but not TLR4, induced migration of DC towards the apical side of the epithelium, demonstrating that there are differences between distinct inflammatory stimuli and the pathways involved in directing DC migration at the local site of inflammation. These experimental set-ups provide new possibilities investigating mechanisms of tissue communication circuits underlying DC motility in the microenvironment [67] of human lung tissue. This is also in line with previous studies that have shown that DC migration is largely influenced by tissue origin, degree of maturity and the 3D structure of the microenvironment [68]. Understanding mechanisms behind changes in DC motility and the imprinting of distinct migratory patterns in response to particular stimulating agents is important in understanding mechanistic actions of infections. Overall these data demonstrate that this protocol is useful for visualizing and dissecting interactions between stromal cells and leukocytes with time-lapse microscopy in models mimicking real tissue and that it is amendable to use with multiple stimuli.

8.5   Future Perspectives

Robust and reliable tissue model systems are key to further advance the field of basic human immunology, as well as the translational exploitation of such basic knowledge. Assays of human stromal cells and leukocyte behaviour in live tissue are generally difficult to perform, but this may be overcome by the use of organotypic model systems that resemble the morphological and functional features of their in vivo parental tissues. Research reviewed herein implicates the usefulness of 3D tissue models to gain further understanding of human tissue immunology in general and tissue-specific cell-leukocyte interactions in particular. Many questions remain to be answered with respect to the contribution of stromal cell and phagocyte cell to tissue homeostasis, inflammation and pathology. The model systems should be amendable to the manipulation of tissue-specific cell gene expression and allow for intervention in combination with monitoring over time using single- and multiphoton imaging. Future models will most likely be more complex and include additional leukocyte populations, and this will enable the design of human tissue models in which we can study both innate and adaptive immune cell components that interact with the stroma. We also foresee that both the tissue-specific cell and leukocyte components can be derived from either healthy individuals or patients. In this context, we recently initiated studies using the skin and lung organotypic models to investigate the functionality of monocytes from Langerhans cell histiocytosis patients, and alveolar macrophages from sarcoidosis patients, respectively. These novel approaches may play a significant role in identifying pathways associated with human disease and may also have great potential in some places replacing experiments performed on animals, for example, in the context of toxicity assays. Increased understanding of the principles and mechanisms of tissue inflammation and pathology using organotypic models can provide strong incentives to initiate new lines of research to generate new treatment regimens and monitoring of disease.

Acknowledgement Our work is supported by grants from the Karolinska Institutet, Stockholm County Council, the Swedish Research Council and Knut and Alice Wallenberg Foundation. This study was, in part, performed at the Live Cell Imaging Unit, Department of Biosciences and Nutrition, Karolinska Institutet.

References

1. Kleinman HK, Philp D, Hoffman MP. Role of the extracellular matrix in morphogenesis. Curr Opin Biotechnol. 2003;14:526–32.
2. Svensson M, Maroof A, Ato M, et al. Stromal cells direct local differentiation of regulatory dendritic cells. Immunity. 2004;21:805–16.
3. Wrenshall L. Role of the microenvironment in immune responses to transplantation. Springer Semin Immunopathol. 2003;25:199–213.
4. Gray TE, Guzman K, Davis CW, et al. Mucociliary differentiation of serially passaged normal human tracheobronchial epithelial cells. Am J Respir Cell Mol Biol. 1996;14:104–12.
5. Roskelley CD, Desprez PY, Bissell MJ. Extracellular matrix-dependent tissue-specific gene expression in mammary epithelial cells requires both physical and biochemical signal transduction. Proc Natl Acad Sci U S A. 1994;91:12378–82.
6. Pampaloni F, Reynaud EG, Stelzer EH. The third dimension bridges the gap between cell culture and live tissue. Nat Rev Mol Cell Biol. 2007;8:839–45.
7. Uller L, Leino M, Bedke N, et al. Double-stranded RNA induces disproportionate expression of thymic stromal lymphopoietin versus interferon-beta in bronchial epithelial cells from donors with asthma. Thorax. 2010;65:626–32.
8. Pflicke H, Sixt M. Preformed portals facilitate dendritic cell entry into afferent lymphatic vessels. J Exp Med. 2009;206:2925–35.
9. Fantin A, Vieira JM, Gestri G, et al. Tissue macrophages act as cellular chaperones for vascular anastomosis downstream of VEGF-mediated endothelial tip cell induction. Blood. 2010;116:829–40.

10. Black AF, Bouez C, Perrier E, et al. Optimization and characterization of an engineered human skin equivalent. Tissue Eng. 2005;11:723–33.

11. Choe MM, Sporn PH, Swartz MA. Extracellular matrix remodeling by dynamic strain in a three-dimensional tissue-engineered human airway wall model. Am J Respir Cell Mol Biol. 2006;35:306–13.

12. Choe MM, Tomei AA, Swartz MA. Physiological 3D tissue model of the airway wall and mucosa. Nat Protoc. 2006;1:357–62.

13. Cukierman E, Pankov R, Stevens DR, et al. Taking cell-matrix adhesions to the third dimension. Science. 2001;294:1708–12.

14. Mishra DK, Sakamoto JH, Thrall MJ, et al. Human lung cancer cells grown in an ex vivo 3D lung model produce matrix metalloproteinases not produced in 2D culture. PLoS One. 2012;7:e45308.

15. Belair DG, Whisler JA, Valdez J, et al. Human Vascular Tissue Models Formed from Human Induced Pluripotent Stem Cell Derived Endothelial Cells. Stem Cell Rev. 2015;11:511–25.

16. Itoh M, Umegaki-Arao N, Guo Z, et al. Generation of 3D skin equivalents fully reconstituted from human induced pluripotent stem cells (iPSCs). PLoS One. 2013;8:e77673.

17. Nguyen Hoang AT, Chen P, Juarez J, et al. Dendritic cell functional properties in a three-dimensional tissue model of human lung mucosa. Am J Physiol Lung Cell Mol Physiol. 2012;302:L226–37.

18. Dezutter-Dambuyant C, Black A, Bechetoille N, et al. Evolutive skin reconstructions: from the dermal collagen-glycosaminoglycan-chitosane substrate to an immunocompetent reconstructed skin. Biomed Mater Eng. 2006;16:S85–94.

19. Dongari-Bagtzoglou A, Kashleva H. Development of a highly reproducible three-dimensional organotypic model of the oral mucosa. Nat Protoc. 2006;1:2012–8.

20. DuMont AL, Yoong P, Surewaard BG, et al. *Staphylococcus aureus* elaborates leukocidin AB to mediate escape from within human neutrophils. Infect Immun. 2013;81:1830–41.

21. Spaan AN, Henry T, van Rooijen WJ, et al. The staphylococcal toxin Panton-Valentine Leukocidin targets human C5a receptors. Cell Host Microbe. 2013;13:584–94.

22. Hieshima K, Imai T, Baba M, et al. A novel human CC chemokine PARC that is most homologous to macrophage-inflammatory protein-1 alpha/LD78 alpha and chemotactic for T lymphocytes, but not for monocytes. J Immunol. 1997;159:1140–9.

23. Russo RC, Garcia CC, Teixeira MM, et al. The CXCL8/IL-8 chemokine family and its receptors in inflammatory diseases. Expert Rev Clin Immunol. 2014;10:593–619.

24. Schutyser E, Richmond A, Van Damme J. Involvement of CC chemokine ligand 18 (CCL18) in normal and pathological processes. J Leukoc Biol. 2005;78:14–26.

25. Birkness KA, Deslauriers M, Bartlett JH, et al. An in vitro tissue culture bilayer model to examine early events in *Mycobacterium tuberculosis* infection. Infect Immun. 1999;67:653–8.

26. Cozens AL, Yezzi MJ, Kunzelmann K, et al. CFTR expression and chloride secretion in polarized immortal human bronchial epithelial cells. Am J Respir Cell Mol Biol. 1994;10:38–47.

27. Schmeichel KL, Bissell MJ. Modeling tissue-specific signaling and organ function in three dimensions. J Cell Sci. 2003;116:2377–88.

28. Khetani SR, Bhatia SN. Microscale culture of human liver cells for drug development. Nat Biotechnol. 2008;26:120–6.

29. Tomasek JJ, Gabbiani G, Hinz B, et al. Myofibroblasts and mechano-regulation of connective tissue remodelling. Nat Rev Mol Cell Biol. 2002;3:349–63.

30. Izumi K, Terashi H, Marcelo CL, et al. Development and characterization of a tissue-engineered human oral mucosa equivalent produced in a serum-free culture system. J Dent Res. 2000;79:798–805.

31. Popov L, Kovalski J, Grandi G, et al. Three-dimensional human skin models to understand *Staphylococcus aureus* skin colonization and infection. Front Immunol. 2014;5:41.

32. Griffith LG, Swartz MA. Capturing complex 3D tissue physiology in vitro. Nat Rev Mol Cell Biol. 2006;7:211–24.

33. Heijink IH, Brandenburg SM, Noordhoek JA, et al. Characterisation of cell adhesion in airway epithelial cell types using electric cell-substrate impedance sensing. Eur Respir J. 2010;35:894–903.

34. Bayless KJ, Kwak HI, Su SC. Investigating endothelial invasion and sprouting behavior in three-dimensional collagen matrices. Nat Protoc. 2009;4:1888–98.
35. Sacharidou A, Koh W, Stratman AN, et al. Endothelial lumen signaling complexes control 3D matrix-specific tubulogenesis through interdependent Cdc42- and MT1-MMP-mediated events. Blood. 2010;115:5259–69.
36. Wilson HK, Canfield SG, Shusta EV, et al. Concise review: tissue-specific microvascular endothelial cells derived from human pluripotent stem cells. Stem Cells. 2014;32:3037–45.
37. Song JJ, Ott HC. Bioartificial lung engineering. Am J Transplant. 2012;12:283–8.
38. Harrington H, Cato P, Salazar F, et al. Immunocompetent 3D Model of Human Upper Airway for Disease Modeling and In Vitro Drug Evaluation. Mol Pharm. 2014 Jul 7;11:2082–91.
39. Nguyen Hoang AT, Chen P, Bjornfot S, et al. Technical advance: live-imaging analysis of human dendritic cell migrating behavior under the influence of immune-stimulating reagents in an organotypic model of lung. J Leukoc Biol. 2014;96:481–9.
40. Nguyen Hoang AT, Chen P, Juaréz J, et al. Dendritic cell functional properties in a three-dimensional tissue model of human lung mucosa. Am J Physiol Lung Cell Mol Physiol. 2012;302:226–37.
41. Parasa VR, Rahman MJ, Ngyuen Hoang AT, et al. Modeling Mycobacterium tuberculosis early granuloma formation in experimental human lung tissue. Dis Model Mech. 2014;7:281–8.
42. Minoo P, King RJ. Epithelial-mesenchymal interactions in lung development. Annu Rev Physiol. 1994;56:13–45.
43. Myerburg MM, Latoche JD, McKenna EE, et al. Hepatocyte growth factor and other fibroblast secretions modulate the phenotype of human bronchial epithelial cells. Am J Physiol Lung Cell Mol Physiol. 2007;292:L1352–60.
44. Banchereau J, Steinman RM. Dendritic cells and the control of immunity. Nature. 1998;392:245–52.
45. Guilliams M, Lambrecht BN, Hammad H. Division of labor between lung dendritic cells and macrophages in the defense against pulmonary infections. Mucosal Immunol. 2013;6:464–73.
46. Pluddemann A, Mukhopadhyay S, Gordon S. Innate immunity to intracellular pathogens: macrophage receptors and responses to microbial entry. Immunol Rev. 2011;240:11–24.
47. Hammad H, Lambrecht BN. Dendritic cells and epithelial cells: linking innate and adaptive immunity in asthma. Nat Rev Immunol. 2008;8:193–204.
48. Lambrecht BN, Hammad H. Biology of lung dendritic cells at the origin of asthma. Immunity. 2009;31:412–24.
49. Godiska R, Chantry D, Raport CJ, et al. Human macrophage-derived chemokine (MDC), a novel chemoattractant for monocytes, monocyte-derived dendritic cells, and natural killer cells. J Exp Med. 1997;185:1595–604.
50. Imai T, Yoshida T, Baba M, et al. Molecular cloning of a novel T cell-directed CC chemokine expressed in thymus by signal sequence trap using Epstein-Barr virus vector. J Biol Chem. 1996;271:21514–21.
51. Pichavant M, Taront S, Jeannin P, et al. Impact of bronchial epithelium on dendritic cell migration and function: modulation by the bacterial motif KpOmpA. J Immunol. 2006;177(9):5912.
52. Loffler B, Hussain M, Grundmeier M, et al. *Staphylococcus aureus* panton-valentine leukocidin is a very potent cytotoxic factor for human neutrophils. PLoS Pathog. 2010;6:e1000715.
53. Puissegur MP, Botanch C, Duteyrat JL, et al. An in vitro dual model of mycobacterial granulomas to investigate the molecular interactions between mycobacteria and human host cells. Cell Microbiol. 2004;6:423–33.
54. Kapoor N, Pawar S, Sirakova TD, et al. Human granuloma in vitro model, for TB dormancy and resuscitation. PLoS One. 2013;8:e53657.
55. Davis JM, Ramakrishnan L. The role of the granuloma in expansion and dissemination of early tuberculous infection. Cell. 2009;136:37–49.
56. Stoop EJ, Schipper T, Huber SK, et al. Zebrafish embryo screen for mycobacterial genes involved in the initiation of granuloma formation reveals a newly identified ESX-1 component. Dis Model Mech. 2011;4:526–36.
57. Volkman HE, Clay H, Beery D, et al. Tuberculous granuloma formation is enhanced by a mycobacterium virulence determinant. PLoS Biol. 2004;2:e367.

58. Mairpady Shambat S, Chen P, Nguyen Hoang AT, et al. Modelling staphylococcal pneumonia in a human 3D lung tissue model system delineates toxin-mediated pathology. Dis Model Mech. 2015;8:1413–25.
59. Inoshima I, Inoshima N, Wilke GA, et al. A *Staphylococcus aureus* pore-forming toxin subverts the activity of ADAM10 to cause lethal infection in mice. Nat Med. 2011;17:1310–4.
60. Wilke GA, Bubeck Wardenburg J. Role of a disintegrin and metalloprotease 10 in *Staphylococcus aureus* alpha-hemolysin-mediated cellular injury. Proc Natl Acad Sci U S A. 2010;107:13473–8.
61. Iwasaki A, Medzhitov R. Control of adaptive immunity by the innate immune system. Nat Immunol. 2015;16:343–53.
62. Lamkanfi M, Dixit VM. Mechanisms and functions of inflammasomes. Cell. 2014;157:1013–22.
63. ONeill LA, Golenbock D, Bowie AG. The history of Toll-like receptors – redefining innate immunity. Nat Rev Immunol. 2013;13:453–60.
64. Iwasaki A. Mucosal dendritic cells. Annu Rev Immunol. 2007;25:381–418.
65. de la Rosa G, Longo N, Rodriguez-Fernandez JL, et al. Migration of human blood dendritic cells across endothelial cell monolayers: adhesion molecules and chemokines involved in subset-specific transmigration. J Leukoc Biol. 2003;73:639–49.
66. Jimenez F, Quinones MP, Martinez HG, et al. CCR2 plays a critical role in dendritic cell maturation: possible role of CCL2 and NF-kappa B. J Immunol. 2010;184:5571–81.
67. Acton SE, Astarita JL, Malhotra D, et al. Podoplanin-rich stromal networks induce dendritic cell motility via activation of the C-type lectin receptor CLEC-2. Immunity. 2012;37:276–89.
68. Gunzer M, Friedl P, Niggemann B, et al. Migration of dendritic cells within 3-D collagen lattices is dependent on tissue origin, state of maturation, and matrix structure and is maintained by proinflammatory cytokines. J Leukoc Biol. 2000;67:622–9.

Index

A
Angiogenesis, 80, 103
Angiostatin, 103
Aorta smooth muscle, 65

B
B-cell activating factor (BAFF), 8, 57
B cell follicle organisation, 8
B cell somatic hypermutation, 16
Blood endothelial cells (BECs), 12, 65
    CCL21, 15
    HEVs, 14, 15
    and LECs, 16
    L-selectin, 15
    proliferation and homeostasis, 15–16
    T and B cells, 14
Blood vascular permeability, 80
BMMSC therapy, 88
Bone remodelling, 44–47
Breast cancer cell invasive behavior, 104

C
CAF-like phenotype, 104
Cancer cell proliferation
    CXCR4, 102
Cancer stemness
    Wnt signaling, 102
Cancer-associated fibroblasts, 117
Carcinoma-associated fibroblasts (CAFs), 101
    antitumoral immune response, 106
    chemotherapy, 104
    ECM, 106
    factors, 107
    fibroblast subsets, 101
    IDO, 107
    inflammatory gene signature, 105
    lysyl oxidase, 103
    markers, 101
    MMP activity, 104
    parabiosis model, 104
    populations, 105
    roles, 102
    STAT1 signaling, 104
    T cell infiltration, 106
Carcinoma-derived cell, 135
Chemotherapy, 104
Chimeric antigen receptor (CAR), 105
Chronic autoimmune disease, 66
Colorectal cancers, 117
Confocal image analysis, 139
Confocal microscopy analyses, 141
Connective tissue growth factor
    (CTGF), 102
Crohn's disease, 88
Cytotoxic effector functions, 77

D
DC migration, 142
Dendritic cells (DCs), 4, 8–9
Dermal fibroblasts, 42
DNA methylation, 43
Double-negative (DN) stroma
    FRCs, 17
    subset, 18
Duchenne muscular dystrophy, 87

E
ECM remodelling factors, 42
Ectopic lymphoid structures, 56
Endothelial cell subsets and leukocytes, 12
Endothelial cells (ECs), 79, 135
Epigenetic mechanisms, 43
Extracellular matrix (ECM), 38, 100

F
Fibroblast
    B and T lymphocytes, 41
    BAFF and APRIL, 42
    DNA methylation, 43
    immune system, 41
    leukocytes, 41
    miRNAs, 44
    MSCs, 38
    and OASF, 42
    PBMCs, 39
    phenotypes, 39, 40
    proinflammatory signals, 40
    proliferation, 40
    pseudoemperipolesis, 42
    RASF and OASF, 44
    stromal address code, 40
    synovial fibroblasts, 43
    TNFα-responsive genes, 41
Fibroblast activation protein (FAP), 40
Fibroblast exosomes, 104
Fibroblastic reticular cells (FRCs)
    B cells, 8
    CD8+ T cell proliferation, 5
    DCs, 4
    deficiency, 30
    deletional tolerance, 7
    interactions, T cells, 7–8
    lymph node, 3
    paracortical T cell zone, 5
    roles, 3
    structural roles, 5
    T cell survival and growth factor, 5
    T zone, 4
    TRA expression, 7
Fibroblasts, 48, 100
Fibroblast-specific protein-1 (FSP-1), 101
Follicular dendritic cells (FDCs)
    B cell follicles, 16
    CXCL13 in lymph nodes, 16

G
Gastrointestinal worms, 25
Germinal centres (GC), 57

Glycoprotein podoplanin, 79
Graft *vs.* host disease (GvHD), 87
Graft-versus-host disease (GVHD) models, 8

H
Haematopoietic cells, 62–63
Haematopoietic stem cell (HSC), 76
Healing wound angiogenesis, 101
High endothelial venules (HEVs), 4, 14
Highly active antiretroviral therapy
        (HAART), 29
Histone acetylation, 43
Histone deacetylase inhibitors, 47
Histone deacetylases (HDACs), 43
HIV infection, 26
Host-directed therapy, 32
Human 3D tissue models, 133–134
Human organotypic models, 136
Human tissue models, 140

I
IFNβ-dependent method, 42
Immune tolerance, 120
Immune-deficient mouse, 103
Indoleamine 2,3-dioxygenase (IDO), 107
Intestinal stromal cells
    adipocytes, 118
    antigen-presenting function, 116
    BMMSC, 117
    CD80 expression, 122
    CD86 expression, 122
    molecule-mediated suppression, 121
    origin, 116–118
    suppression, soluble mediators, 122
    T cell proliferation, 122
    tMSC, 117
Intratumoral T cells, 107

L
Leishmania donovani infection, 28
Leukocyte component, 135–136
Leukocyte interaction, 139
Leukocyte recruitment, 81–82
Live-imaging 3D tracking, 140
Live-imaging experiments, 137
LTβR and TNF-RI-signalling pathways, 65
Lung tissue microenvironment, 138
Lung tissue models, 138
Lymph node stromal cells (LNSCs), 2, 6, 28
Lymph nodes, 1
Lymphatic endothelial cells (LECs), 12

ACKR2, 11
   deletional tolerance, 13–14
   structural and chemoattractive role, 11
   suppressive tolerance, 13
   T cells, 13
Lymphatic vessels, 65
Lymphocytes, 77
Lymphoid tissue inducer (LTi) cells, 10
Lymphoid tissue microenvironment, 27
Lymphotoxin-β receptor (LTβR) signalling, 58

M
Marginal reticular cells (MRCs), 9, 63
   B cell chemoattractant, 10
   cytokines IL-7 and RANKL, 10
   FDCs, 10
   lymphoid function and SLO, 11
   organogenesis, 10
   SCS, 10
MCMV infection model, 26
Mesenchymal stem cell replacement, 118
Mesenchymal stem cells (MSCs)
   acute inflammation, 83–86
   adaptive immune system, 78
   adaptive immunity, 78–79
   adipocytes and osteoblasts, 87
   angiogenesis, 80
   behaviour, 86
   β-catenin, 76
   BM-derived, 75
   BMMSC and WJMSC infusion, 81
   BMMSC, 76, 77
   BM-resident, 77
   clinical benefits, 88
   CXCL12-CXCR4-dependent, 76
   cytokines, 83
   definition, 75
   and EC, 81
   and endothelial cells, 81
   GvHD patients, 88
   HLA expression, 78
   HSC niche, 76
   innate immunity, 77–78
   leukocyte recruitment, 81–82
   molecule-mediated suppression, 118
   neutrophils, 77
   origin, 75
   platelets, 79
   populations, 82
   SLE, 86
   soluble mediator-mediated suppression, 121
   systemic infection, 76
   T-cell behaviour, 78
   tissue-resident MSC, 83
   TLR signaling, 121
   treatment, 89
Microarray analysis, 39
MicroRNAs (miRNAs), 44
Molecule-mediated suppression, 118
Mucosal stromal cells, 116
Myeloid differentiation factor 88
     (MyD88), 123
Myofibroblast/fibroblasts, 117

N
Natural disease cycle, 88
Neutrophils, 77
Neutrophils and lymphocytes, 81
Non-haematopoietic cells, 2, 63–65
Non-haematopoietic lymph node stromal
     cells, 17
Non-haematopoietic stromal cells, 16
Nonobese diabetic (NOD), 58

O
Organotypic culture models, 132
Organotypic models, 135
Osteoblast-mediated bone formation, 45
Osteoblasts, 45
Osteoclast-mediated bone resorption, 45

P
Parenchymal mesodermal cells, 116
PD-L1 and PD-L2, 116, 122, 123
Peripheral blood mononuclear cells
     (PBMCs), 38
Peripheral tissue antigen (PTA), 7
Phagocyte cell migration, 139, 142
Phenotypic comparison, 119–120
Physical microenvironment, 83
Platelet-derived growth factor (PDGF), 102
Platelet-MSC interactions, 79
Podoplanin, 2, 5, 8, 11, 17–18
Prostaglandin secretion, 117
Prostate fibroblasts, 102
Pseudomonas aeruginosa infection, 62

R
RA synovial tissue, 44
Rheumatoid arthritis (RA)
   bone maintenance, 44–45
   bone remodelling, 44–47
   environment, 44

S
Salivary duct epithelium, 65
Sjögren's syndrome (SS), 57
Streptococcus pyogenes infections, 100
Stroma, 134–135
Stromal and tumor immune responses, 106
Stromal architecture, 30
Stromal cell biology, 28
Stromal cell tropism, 25
Stromal cell-leukocyte interactions
    biological material, 135
    DC, 138
    donor variability, 132
    fibroblast component, 134
    human 3D tissue models, 133–134
    leukocyte component, 135–136
    pathogens and inflammatory reactions, 132
    stroma, 134–135
    3D tissue models, 132
    tissue microenvironment, 139–140
Stromal cells
    angiogenesis, 103
    CAFs, 101–102
    cancer cell proliferation, 102
    cancer stemness, 102–103
    fibroblasts, 100–101
    intratumoral T cells, 107
    invasion and metastasis, 103–104
    tumor microenvironment, 100
Stromal infection
    animal models, 31
    chronic inflammation, 31
    DC, 29
    DC and T cell, 30
    denuded marginal zone, 31
    ectopic lymphoid tissue, 31
    and epithelial cells, 24
    FDC, 29
    fibrocytes, 26
    HAART, 29
    haematopoietic, 26
    HIV and visceral leishmaniasis, 30
    HIV infection, 26
    HSV, 25
    immune system, 29
    innate immunity, 26
    LCMV infection, 30
    lymph nodes, 29
    lymphocyte and dendritic cell functions, 29
    lymphoid tissues, 27
    MHCII-peptide, 27
    mucosal pathogens, 25
    NLRC4 inflammasome, 26
    pathogen cellular tropism, 26
    pathogens, 24, 25
    role, 23
    T cells and DC, 30
    TB granulomas, 31
    temporal role, 24
    vector-borne parasites, 25
Subcutaneous tumour apoptosis, 65
Sublining fibroblasts, 40
Synovial fibroblasts, 40–42
Systemic lupus erythematosus (SLE), 86

T
T cell infiltration, 106
T cell recruitment, 108
T cell tolerance induction, 7
Tartrate-resistant acid phosphatase
        (TRAP), 46
Teratocarcinoma cells, 100
Tertiary lymphoid structures (TLSs)
    B-cell activation, 57
    chemokines, 58–60
    cytokines, 61–62
    definition, 56–57
    fibroblastic stromal cells, 63
    function, 57–58
    haematopoietic cells, 62–63
    immune activity and persistence, 58
    immune cells, 56
    LTαβ or TNFα, 64
    LTβ-Fc, 65
    LTβR, 64
    lymphatic endothelial cells, 56
    lymphedema, 56
    molecular cues, 59–62
    non-haematopoietic cells, 63–65
    pathologies, 65
    RA, 58
    SLOs, 56
    spontaneous *vs.* induced, 58–59
    SS, 57
    T- and B-cell survival factors, 57
    TNFSF members, 60–61, 64
    transient formation, 57
    treatments, 65
3D microenvironment, 142
3D rendering of tissue model, 138
Thymic stromal lymphopoietin
        (TSLP), 27
Tissue bacterial toxin-induced
        pathology, 141
Tissue inflammation, 143
Tissue inhibitors of metalloproteinases
        (TIMPs), 104

Tissue model set-up, 137
Tissue repair, 82
Tissue-resident MSC, 82
Tissue-resident stroma, 79–82
Tissue-restricted antigens (TRAs), 7
TNF-related apoptosis-inducing ligand
        (TRAIL), 47
Tolerogenic responses, 123
Toll-like receptors (TLRs), 40, 142
Treg cells, 108
Tuberculosis (TB), 140
Tumor immunology, 105
Tumor microenvironment, 100
Tumor-associated macrophages (TAMs), 102

V
Vascular endothelial cells, 79–82
Vascular endothelial growth factors (VEGF),
        103
Vesicular stomatitis virus (VSV) infection, 26

W
Wnt signaling, 102
Wound healing macrophages, 100

X
Xenograft model, 103

The manufacturer's authorised representative in the EU is Springer
Nature Customer Service Centre GmbH, Europaplatz 3, 69115 Heidelberg,
Germany. If you have any concerns regarding our products, please
contact ProductSafety@springernature.com

Printed and bound by CPI Group (UK) Ltd, Croydon, CR0 4YY
28/11/2025
02007732-0006